University of Delaware Solar House, built in 1973, uses flat plate collectors incorporating cadmium sulphide solar cells to provide heat and electric power. Lead acid batteries are used for electrical storage. The house uses dc current for the stove, lights and some appliances, and an invertor provides ac power to the refrigerator, heat pump and fans. (Photo courtesy of Karl Böer.) See page 12.

SOLAR ENERGY
Technology
and
Applications

SOLAR ENERGY
Technology
and
Applications

by
J. Richard Williams, Ph.D.
Associate Professor of Mechanical Engineering
Georgia Institute of Technology
Atlanta, Georgia

Geology

PREFACE

Solar energy represents the only totally non-polluting inexhaustible energy resource that can be utilized economically to supply Man's energy needs for all time. Recent large price increases and reduced availability of fossil fuels, and public concern about the safety of nuclear reactors have led to a surge of interest in using the power of the sun.

Solar Energy Technology and Applications introduces the various techniques for utilizing solar energy and brings you up to date on work to the present time on the broad spectrum of solar energy systems. Succinctly written for the technical person, it is readily understood by anyone with no background in the field. It is also recommended as a supplementary text for energy-related courses.

ACKNOWLEDGMENTS

I would like to express my appreciation for the support of this research by the National Aeronautics and Space Administration during the past eight years. Also the guidance and assistance of Robert Ragsdale and George Kaplan of the Lewis Research Center is gratefully acknowledged.

J. Richard Williams

CONTENTS

Chapter 1

SOLAR ENERGY:
ITS TIME HAS COME

In 1970 the total energy consumed in the United States was about 65×10^{15} BTU, which is equal to the energy of sunlight received by 4300 square miles of land, or only 0.15% of the land area of the continental U.S.[1] Even if this energy were utilized with an efficiency of only 10%, the total energy needs of the U.S. could be supplied by solar collectors covering only 1.5% of the land area, and this energy would be supplied without any environmental pollution. With the same 10% utilization efficiency, about 4% of the land area could supply all the energy needs in the year 2000. By comparison, at present 15% of the U.S. land area is used for growing farm crops.[2] For some applications, such as heating water and space heating for buildings, the utilization efficiency can be much greater than 10%, and the collectors located on vertical walls and rooftops of buildings, so the 4% estimate represents an upper limit and actual land area requirements can be considerably smaller.

As a practical matter, even though sunlight can provide all our energy needs without pollution, in the foreseeable future solar energy will not provide all or even most of this energy. Over the past century fossil fuels have provided most of our energy because energy from fossil fuels has usually been cheaper and more convenient than energy from available alternative energy sources, and until recently environmental pollution has been of little concern. The construction of large nuclear electric generating plants is presently underway, and nuclear power will play an increasingly important role; so in the coming decades a variety of energy sources

1

will supply the U.S. energy needs, and solar energy will only be utilized when it is competitive with alternative energy sources.

Over the past few years energy forecasts[3-12] have been made which predict large increases in the consumption of oil and coal as well as a rapid increase in nuclear generation. However, these forecasts predict that the domestic production of oil would not be sufficient to keep pace with demand, so large increases in oil imports would be necessary. The recent rapid escalation of the cost of foreign crude oil has cast doubt on the ability of the U.S. to supplement its energy needs from foreign imports, so the President has urged that the U.S. become self sufficient in its energy supplies. This will require developing additional domestic energy resources. Solar energy, which so far has seen insignificant use in the U.S., can be utilized to make a significant impact as a new energy resource over the next few years. The most immediate large-scale applications would be for the heating and cooling of buildings, heating water, and supplying heat for industrial and agricultural drying operations. Over the longer term, solar energy can also be used for pollutionless electric power generation.

The NSF/NASA Solar Energy Panel[13] identified three broad applications as "most promising from technical, economic, and energy quantity standpoints. These are: (1) the heating and cooling of residential and commercial buildings, (2) the chemical and biological conversion of organic materials to liquid, solid, and gaseous fuels, and (3) the generation of electricity". It also reported that "solar energy can be developed to meet sizable portions of the Nation's future energy needs". Energy for space heating, air conditioning, and water heating for buildings presently accounts for about 25% of the total energy consumption in the U.S.,[14] and virtually all this energy is supplied by the combustion of high quality fossil fuels. Sunlight can provide about half this energy at a cost competitive with fossil fuels, thus reducing total consumption of fossil fuels by more than 12%.[15-16] In the future sunlight can also be utilized directly for electric power generation and for producing fuels to replace the fossil fuels now being used as an energy source. Coal, oil, and gas are non-renewable resources of great value for producing fabrics, plastics and other materials, and probably should not be burned once economical alternative fuels become available.

Chapter 2

AVAILABILITY OF SOLAR ENERGY

To evaluate the economics and performance of systems for the utilization of solar energy in a particular location, a knowledge of the available solar radiation at that place is essential. Thus, the utilization of solar energy, as with any other natural resource, requires detailed information on availability.

For approximate calculations, average values of energy availability are often used. Solar energy arrives at the surface of the United States at an average rate of about 1500 BTU/ft^2/day (about 42×10^9 BTU/mi^2/day). Over the period of a year a square mile receives about 15×10^{12} BTU. In 1970 the total energy consumed by the U.S. for all purposes was about 6.5×10^{16} BTU. Thus 4300 sq mi of continental U.S. land receives on the average in one year the equivalent of all the U.S. energy needs.[2] At 10% conversion efficiency, 43,000 square miles (about 1.5% of the land area of the 48 contiguous states) could produce the amount of energy the United States consumed in 1970. Solar energy availability has been described as a double periodic function with a 24 hour and a 365 day period length, super-imposed with a fluctuating screening function (cloud over).[17] The maximum amplitude of this function is approximately 300 BTU/ft^2, and for the continental U.S.A. (excluding Alaska) it integrates to an average influx of approximately 570,000 BTU/ft^2/year. Thus, the yearly 24 hour average solar energy received in the United States is about 65 BTU/ft^2 hr.

However, precise evaluation of proposed solar energy systems requires accurate data on the solar intensity, spectrum, incident angle, and cloudiness as a function of time at the place where the solar energy system is to be located. Past surveys of worldwide solar radiation have been based on very limited data for most areas. A large amount of data is available in the United States and Japan on the time-dependent direct and diffuse intensity function, but many solar applications require data on the probability of cloudy periods of specific duration, and this type of data is seldom available. Also, in some cases the results of radiation surveys are reported on an annual basis only, which precludes the use of this information for the rational design of solar energy systems in most areas where seasonal variations of radiation are large.

Several types of solar radiation data are available including direct radiation at normal incidence, direct plus diffuse radiation at normal incidence, direct radiation on a horizontal surface, direct plus diffuse radiation on a horizontal surface, and each of these on tilted and vertical surfaces.

For each type of measurement, one may wish to know the maximum and minimum values in selected time periods. It is also necessary to decide what averaging should be employed—seasonal, monthly, daily, hourly, or even shorter intervals. For devices employing focusing systems, data on the intensity and direction of direct radiation would be required. For flat-plate collectors the total (direct plus diffuse) radiation intensity on a sloping surface is needed for collectors used in that position. Maximum radiation values are needed to determine the capacity of solar energy systems; evaluation of the system performance over an extended period of time requires data on average intensity and appropriate time-intensity distribution parameters. The form of data most generally used are obtained by continuous monitoring stations which record direct and diffuse solar radiation intensity in a form compatible with digital computers. The form of the data most available and most frequently reported is total radiation (direct plus diffuse) on a horizontal surface received each day or in some cases each hour. Approximate methods are available for estimating the direct component and distribution parameters from the total radiation data.

Solar radiation (referred to technically as solar insolation, or just insolation) is measured by several different types of instruments having various characteristics and degrees of accuracy. Thermoelectric pyranometers, introduced by Kimball and Hobbs in 1923, measure the difference in temperature between black and white surfaces in a glass-enclosed chamber due to the differences in solar-radiation absorption. The electric current output from the thermopiles in these units is recorded on a chart, magnetic tape, or other instrument. If well calibrated and maintained, these instruments can provide long-term and short-term values of solar and sky radiation with an accuracy of 3% or less.

Another type of instrument utilizes the differential expansion of a bimetallic element due to solar absorption to move the stylus on a strip chart recorder. Its accuracy is typically within about 10%. Another radiation instrument is the Bellani pyranometer, which provides an indication of total solar radiation by the quantity of a liquid that has distilled from a solar-heated evaporating chamber. The amount of liquid distilled during a given period provides a measure of the total insolation received during that period. In the United States and Europe the thermoelectric pyranometer is most frequently used. The bimetallic type is simpler and cheaper, and fairly widely used in South America and Asia, as well as in scattered stations elsewhere in the world.[18]

The number of hours of sunshine per day can be measured by the Campbell-Stokes sunshine recorder which uses a lens to focus direct sunshine onto a heat-sensitive paper chart. A discolored line is produced when the solar disc is focused onto the paper. The length of the discolored line divided by the total length of the chart corresponding to the time between sunrise and sunset is the percent possible sunshine for the day.

Regular measurements of sunshine duration and cloudiness are made at numerous weather stations throughout the world, and these records usually cover periods of 20 to 60 years or more. The average daily radiation is a function of sunshine duration at the particular location, and is correlated with the amount received outside the atmosphere Q_0 by

$$Q = Q_0 \left(a + b \frac{S}{S_0} \right)$$

where Q is the average daily radiation received at the surface location, S is the number of hours of sunshine recorded at the site per day, and S_o is the maximum possible number of hours of sunshine at the site per day (unobstructed horizon), and a and b are constants (Table 1).[19] The average value of Q_o is equal to 429 BTU/hr/ft^2 multiplied by the cosine of the latitude angle.

Table 1. Climatic Constants[18]

Location	S/S_o	a	b
Charleston, S.C.	0.67	0.48	0.09
Atlanta, Ga.	0.59	0.38	0.26
Miami, Fla.	0.65	0.42	0.22
Madison, Wisc.	0.58	0.30	0.34
El Paso, Texas	0.84	0.54	0.20
Poona, India (Monsoon)	0.37	0.30	0.51
(Dry)	0.81	0.41	0.34
Albuquerque, N.M.	0.78	0.41	0.37
Malange, Angola	0.58	0.34	0.34
Hamburg, Germany	0.36	0.22	0.57
Ely, Nevada	0.77	0.54	0.18
Brownsville, Texas	0.62	0.35	0.31
Tamanrasset, Sahara	0.83	0.30	0.43
Honolulu, Hawaii	0.65	0.14	0.73
Blue Hill, Mass.	0.52	0.22	0.50
Buenos Aires, Arg.	0.59	0.26	0.50
Nice, France	0.61	0.17	0.63
Darien, Manchuria	0.67	0.36	0.23
Stanleyville, Congo	0.48	0.28	0.39

The United States National Weather Service solar radiation network presently has over 90 measuring sites. A few of these use Eppley Model II pyranometers which have about twice the accuracy of instruments typically used at the other sites. Data are stored at one minute intervals on magnetic tape (which can later be processed by computer). The various primary standards for solar radiation that have been used have been shown to differ from each other as much as 6%, so care must be taken in comparing data from different instruments and from different sites. Instruments which are being used may degrade as much as 20%–30% before being replaced, so measured intensities can be 20%–30% low for this reason. Some sites, however, have very good data with an accuracy of 2–3%. Much of the data is available from the National Weather Service as hourly data on tape or cards, and a data format manual is also available.

Hourly or daily data are no longer published in printed form at the national level, but only in card, tape or microfilm form. Differences between monthly average sunshine may differ about 40% from year to year and typically 20%–30% from site to site. There may be large differences between nearby sites due to local weather differences, and there can be sizable differences from year to year because of changes in atmospheric turbidity.

Efforts are presently underway to relate reflected solar radiation to ground level incident radiation so that satellite measurements can be made useful for terrestrial solar energy applications. Absolute deviation of measurements of the solar constant versus wavelength is less than 5% using spectral radiometers.

Surface albedo is determined by taking the 15-day minimum value of reflected sunlight measured by the satellite, and once this value is determined, it can be used to evaluate incident surface radiation from satellite measurements. Satellite measurements should provide very useful data over short time scales, but cannot be extrapolated over long time periods because of variations in surface albedo and atmospheric turbidity. Satellite measurements are needed for microscale data (resolution a few miles). Since interpolation between weather stations is not adequate for specific site studies of solar-thermal conversion this data must either come from satellites or from a continuous insolation monitoring station at the site. One problem, however, is that satellites provide data on total radiation, whereas for systems using solar concentrators, direct beam radiation is needed. One can determine this if the cloudiness is measured, and satellites do measure cloudiness. Recent measurements by NASA determined the solar constant to be 429.0 ± 0.5 BTU/hr·ft^2 (1353 ± 1.5 W/m^2) outside the atmosphere.[20]

The flat plate collector incorporates a transparent cover over a black plate with air or water flowing over or through the black plate, and is usually fixed in position. To evaluate its performance, one must know the intensity, angle and spectrum of solar energy as a function of time. Surface reflectivities depend on the incidence angle, and incident radiation must be split into direct and diffuse components. Empirical techniques have been developed for doing this by using a relationship between daily total radiation outside the atmosphere and daily total at ground level. Statistical distribution curves of hourly radiation versus fraction of time radiation received are very similar for different sites of equivalent overall cloudiness. These data and data on the probability of two consecutive days of cloudiness, are needed to determine energy storage requirements of proposed solar energy systems.

One typical insolation measuring station which provides the necessary data for the evaluation of experimental solar energy projects consists of the following equipment:

Kimball-Hobbs thermoelectric pyranometer	$ 560
Spectral pyranometer	990
Normal incidence pyrheliometer	880
Equatorial mount	1225
Wind and temperature instrumentation	1100
Total Cost (exclusive of recorders)	$4755

The prices given are February 1974 prices as quoted by two well known vendors. The Kimball-Hobbs pyranometer continuously monitors total insolation in the wavelength range from 0.285 to 2.8 microns. The spectral pyranometer is a similar instrument with hemispherical filters which allow only insolation in restricted wavelength ranges to reach the black and white surfaces, so total (direct plus diffuse) insolation in the restricted wavelength range is measured. The pyrheliometer is supported on the electrically driven equatorial mount and measures direct beam solar radiation only. The wind instrument consists of a low threshold cup anamometer and a vane for measuring wind speed and direction over a speed range of 0.4–100 mph. Three thermisters are used for recording the ambient temperature at heights of 3 meters and 10 meters. These equipment are presently used at an insolation station established at the Georgia Institute of Technology, Atlanta, and provide the necessary insolation and weather data for evaluating the performance of flat plate collectors and other devices that collect solar energy.

Chapter 3
SOLAR ENERGY COLLECTORS

The type of device used to collect solar energy depends primarily on the application. Flat plate thermal energy collectors are used for heating water and heating buildings, but can provide temperatures of only about 150°F above ambient. If higher temperatures are desired, the sunlight must be concentrated onto the collecting surface. If electrical power is to be produced, photovoltaic cells can be used to convert sunlight directly into electricity, either with or without concentrators. The decision as to what kind of collector should be used for a specific application is dictated by economics.

Flat Plate Collectors

Figure 1 illustrates the basic components of a flat plate collector. A black plate is covered by one or more transparent cover plates of glass or plastic, and the sides and bottom of the box are insulated. Sunlight is transmitted through the transparent covers and absorbed by the black surface beneath. The covers tend to be opaque to infrared radiation from the plate, and also retard convective heat transfer from the plate. Thus, the black plate heats up and in turn heats a fluid flowing under, through, or over the plate. Water is most commonly used, since the temperatures involved are usually below the boiling point of water. The hot water may be used directly or for space heating in homes and buildings. Tests of five types of flat plate collectors using flat black absorber plates showed that the collection efficiency ranged from 40%–60% for a 30°F temperature rise and dropped to 30% or less for a 100°F temperature rise.[21]

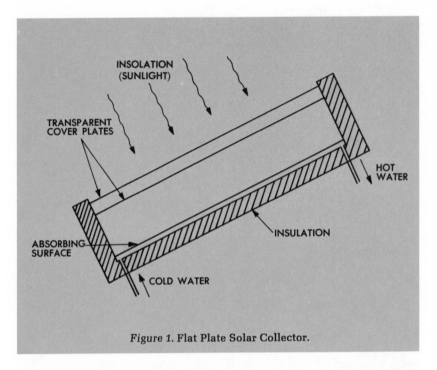

Figure 1. Flat Plate Solar Collector.

These collectors consisted of a wood frame with the flat black absorber inside; 7–10 centimeters of sawdust were used beneath the absorbing surface for insulation, and the top of the frame was covered with a single 2 mm thick window glass. The absorber was a 1 meter by 3 meter steel sheet of 2.2 mm thickness containing 1 cm diameter coolant tubes 10 cm apart. The maximum incident radiation intensity was 800 Kcal/m²/hr (295 BTU/ft²hr).

Tests of collectors consisting of two glass panes and a flat black metallic absorber which studied the effects of varying the air gap between the glass panes and between the glass and collector plate (0.01, 0.02, 0.04 and 0.08 ft) showed that the best performance was with the 0.08 ft spacing, but the performance with the 0.04 ft spacing was almost as good.[22] The performance of these collectors was considerably improved when a selective coating was applied to the collecting surface, instead of flat black paint. Figure 2 illustrates the spectral reflectance of three types of selective coatings.[23] Such coatings strongly absorb incident sunlight, but retard reradiation

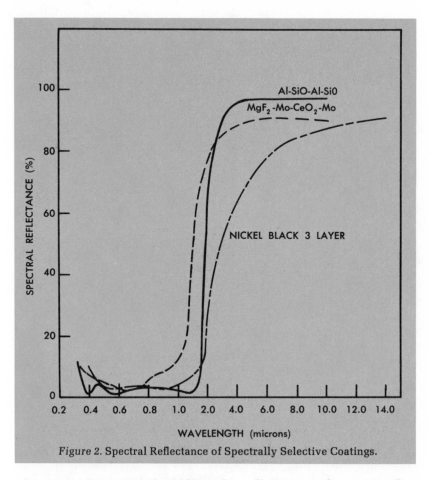

Figure 2. Spectral Reflectance of Spectrally Selective Coatings.

of infrared heat, and thus allow the collecting surface to reach a higher equilibrium temperature. For a 100°F temperature difference between the outer glass and absorber the collection efficiency increased from 35% to 55% when the selective coating was added, and increased from 10% to 40% when the temperature difference was 150°F.[22] However, the cost of the collector was also increased, so there was no major change in its cost effectiveness. The collection efficiency of dual glass plate vertical collectors was measured as a function of temperature for three insolation levels. The maximum temperature difference reached was 87°F for an insolation of 100 BTU/ft² hr, 153°F at 200 BTU/ft² hr, and 210°F at 300

BTU/ft² hr. The collection efficiency was about 50% at half the maximum temperature, and decreased almost linearly to 0 at the maximum temperature.

The efficiency of flat plate collectors can also be improved by anti-reflective coatings on the transparent covers. Figure 3 illus-

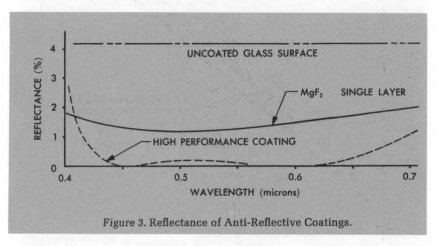

Figure 3. Reflectance of Anti-Reflective Coatings.

trates the percent of normal incidence sunlight reflected from uncoated and coated glass surfaces.[24] Coated surfaces cost more than uncoated surfaces, of course, and the coating cost increases as the perfomance increases.

Figure 4 illustrates a typical flat plate collector used to provide hot water for space heating and the operation of absorption-type air conditioners. Such collectors are placed on rooftops with a southward slope and on south-facing walls. The average daily insolation is reduced about 20% during the winter if the wall faces southeast or southwest instead of south, and is reduced about 60% if the wall faces east or west. A flat plate collector incorporating solar cells has been developed at the University of Delaware[25] (Figure 5) to supply both electricity and heat for a house. One problem with this type of collector is the decrease in photovoltaic conversion efficiency and lifetime with increasing temperature. The 4 × 8 ft collectors are deployed between the roof joists from the inside; the outside is glased with ¼ inch plexiglas. The heat transfer fluid for this type of collector is air.

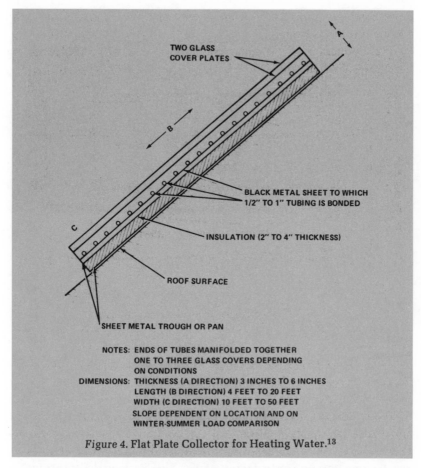

TWO GLASS
COVER PLATES

BLACK METAL SHEET TO WHICH
1/2" TO 1" TUBING IS BONDED

INSULATION (2" TO 4" THICKNESS)

ROOF SURFACE

SHEET METAL TROUGH OR PAN

NOTES: ENDS OF TUBES MANIFOLDED TOGETHER
ONE TO THREE GLASS COVERS DEPENDING
ON CONDITIONS
DIMENSIONS: THICKNESS (A DIRECTION) 3 INCHES TO 6 INCHES
LENGTH (B DIRECTION) 4 FEET TO 20 FEET
WIDTH (C DIRECTION) 10 FEET TO 50 FEET
SLOPE DEPENDENT ON LOCATION AND ON
WINTER-SUMMER LOAD COMPARISON

Figure 4. Flat Plate Collector for Heating Water.[13]

Scientists at NASA's Marshall Space Flight Center have produced a flat plate collector with a new type of selective coating which absorbs over 90% of the incident solar radiation while reemitting only about 6%. A maximum absorber temperature of 450°F was reported without water in the coils.[26]

Solar Concentrators

Concentrators may be used to produce temperatures in excess of 300°F for efficient electrical power generation, for industrial and agricultural drying, and for other applications where high tempera-

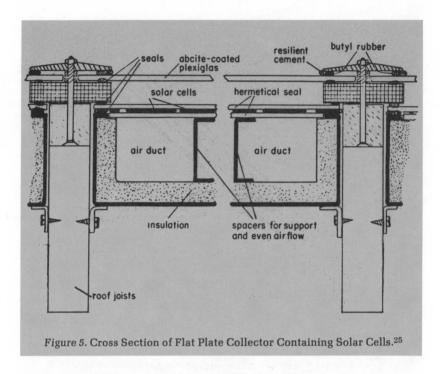

Figure 5. Cross Section of Flat Plate Collector Containing Solar Cells.[25]

ture heat is needed. Also, concentrators have been used to increase the power output of photovoltaic cells.

One method of concentrating the sun's rays is through a lens. A small magnifying glass will concentrate enough heat to burn wood or paper. Making a large enough lens to concentrate a large amount of incident radiation is difficult, so most solar concentrators employ a reflector system. For high concentration the ideal form of the concentrator, from an optical standpoint, is parabolic; however, to achieve this high concentration the reflector must be steered to remain directed toward the sun, and the heat exchanger must remain located at its focus.[27] For this reason, parabolic concentrators are seldom considered for most solar energy applications where the cost of collecting the solar energy must be kept low. Large solar collectors are subject to large wind loadings, and thus require a sturdy supporting structure.

Since the sun has an apparent diameter with an angle of approximately 0.009 radians, the ideal parabola will concentrate most of the radiation on a hot spot with a diameter equal to the focal length times 0.009 radians. The concentration ratio C produced at the hot spot is defined as the ratio of the solar radiation intensity on the hot spot to the unconcentrated direct sunshine intensity at the concentrator site:

$$C = \frac{q_f}{q_i} = \frac{\text{solar radiation intensity at hot spot}}{\text{unconcentrated direct solar radiation}}$$

For a perfect paraboloid the ratio is found to be a function of the paraboloid rim angle θ and the angular diameter of the sun (a = 0.009 radians).

$$C = \frac{4}{a^2} \sin^2 \theta$$

or:

$$q_f = q_i \frac{4}{a^2} \sin^2 \theta$$

The parabola must be oriented with its axis toward the sun. One method of maintaining this orientation is to rotate the parabolic concentrator tracking the sun's motion across the sky, and move it about two axes to account for seasonal as well as daily motion.

An alternative to steering the concentrator is to use auxiliary mirrors. In this approach a large flat plane mirror is used to track the sun and reflect its rays into the parabolic concentrator. The surface of the plane mirror must be of greater area than the paraboloid because of the angles involved, but its structure is less complex than the paraboloid. Because the reflecting surface efficiency of any surface is less than 100%, the use of auxiliary mirrors results in an efficiency loss. The cost factor along with the advantages of a fixed working surface make auxiliary mirrors common in modern parabolic solar concentrators.

For a solar concentrator the maximum temperature obviously cannot exceed the temperature of the sun. Theoretically, for an ideal paraboloid of 100% reflectivity operating in space, the temperature of a black body at the hot spot would reach approximately 10,000°F. However, the atmosphere reduces the incident radiation of the sun. In the United States the ratio of average total incident radiation received to that received outside the atmosphere ranges from over 0.75 to under 0.40. The reflectivity of aluminum ranges from 0.89 to 0.53, and for mirror glass this reflectivity ranges from

0.72 to 0.96. These factors along with other irregularities limit the reported temperatures for existing solar furnaces to about 7000°F. A small parabolic concentrator is capable of producing the same temperature as a large concentrator, the difference being only the amount of heat collected and the size of the hot spot. The hot spot on most moderate to large size furnaces is about 1–2 inches in diameter. To produce a hot spot of five inches diameter requires a reflector with a diameter equal to the height of a 10-story building. Concentrators using toroidal, flat or spherical components are usually cheaper to produce than parabolic concentrators.[28]

One simple type of concentrating solar collector (Figure 6) uses a

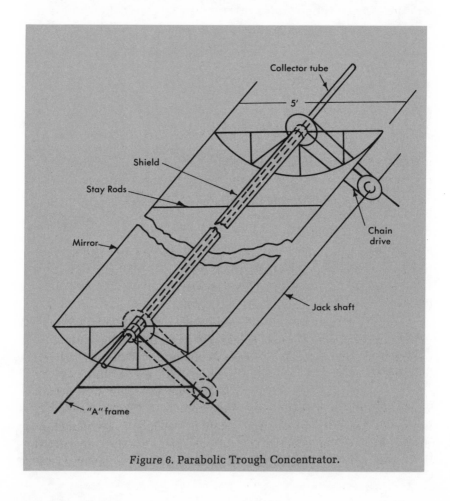

Figure 6. Parabolic Trough Concentrator.

parabolic cylinder reflector to concentrate sunlight onto a collecting pipe within a quartz or pyrex envelope. The pipe can be coated with a selective coating (Figure 3) to retard infrared emission, and the transparent tube surrounding the pipe can be evacuated to reduce convective heat losses. The reflector is steered during the day to keep sunlight focused on the collector. This type of concentrator, known as the parabolic trough concentrator, cannot produce as high a temperature as the parabolic reflector, but produces much higher temperatures than flat plate collectors.

Solar-thermal collectors may be categorized as (1) low temperature flat-plate collectors with no concentration, (2) medium temperature concentrating collectors typified by parabolic cylinders, and (3) high concentration, high temperature collectors such as parabolic concentrators or concentrators composed of many flat mirrors focused at the same point. Table 2 gives the usual temperature ranges and the collection efficiencies for these three categories of collectors.

Table 2. Classification of Solar Collectors

Category	Example	Temperature Range	Efficiency
No Concentration	Flat Plate	150– 300 °F	30–50%
Medium Concentration	Parabolic Cylinder	500–1200 °F	50–70%
High Concentration	Parabodial	1000–4000 °F	60–75%

The actual temperature obtained will depend on the optical performance of the reflector, the accuracy of the tracking device, and the absorption efficiency of the receiver.[29]

Researchers in the Soviet Union have developed a technique for mass producing inexpensive faceted solar concentrators which form an approximate parabolic cylinder. They used a jig containing a number of vacuum socket facet holders, arranged along a convex cylindrical parabolic surface, all connected to a central vacuum system. In making a concentrator, the 26 mirror strips were placed face down on the correctly positioned holders and the vacuum held the mirror facets in the desired position throughout the manufacturing process. The reverse side of the mirrors was then coated with a layer of epoxy resin and covered with glass fabric. The sup-

porting structure, which had the approximate surface shape of the
finished concentrator, was placed on the glass fabric and glued to
the mirror. After the epoxy had cured, the vacuum was turned off
and the finished concentrator removed. Soviet researchers manufactured 80 concentrator sections one meter long by about one
meter wide using this technique. These concentrator sections were
used to make two power plants. It was only necessary to align the
sections, and not the individual facets. These concentrators were
cheap to produce, had good optical characteristics, and were quite
strong.[30]

Two general approaches have been used to try to reduce the expense and engineering difficulties associated with steering the reflecting surface: 1) use simple automatic steering mechanisms to
move many separate reflectors, which require less supporting structure than a single large concentrator, and 2) develop concentrators
with fixed mirrors and movable heat collectors.

One of the most promising passive steering devices for cylindri-
cal-type concentrators and other small collectors is the thermal
heliotrope, which consists of a single bimetallic coil with appropriate thermal coatings and a feedback shade,[31] shown schematically
in Figure 7. The fixed end of the helix is attached to a stationary
support and the solar array and the feedback shade are attached
to the free end of the helix. The shade regulates the amount of solar
radiation incident on the helix, causing the rotation of the helix to
stop when the array is properly aligned.

If θ_o represents the initial angle of the sun, then sunlight falling on
the helix causes its temperature to rise and the two components of
the helix to rotate the solar array toward the sun. At angle θ_A the
shade begins to cast a shadow on the helix, and additional rotation
of the helix causes the shade to shield a portion of the helix from
the solar radiation. This decreased solar energy input reduces the
rate at which the temperature of the helix was increasing; this, in
turn, reduces the rotational velocity of the helix until the net energy
input to the helix is zero and the rotation ceases, and the solar
array is aligned perpendicular with the sun's rays. The bimetallic
helix can also be used to orient cylindrical or parabolic concentrators toward the sun. As the relative position of the sun changes
such that θ_o increases slightly in a clockwise direction, the surface

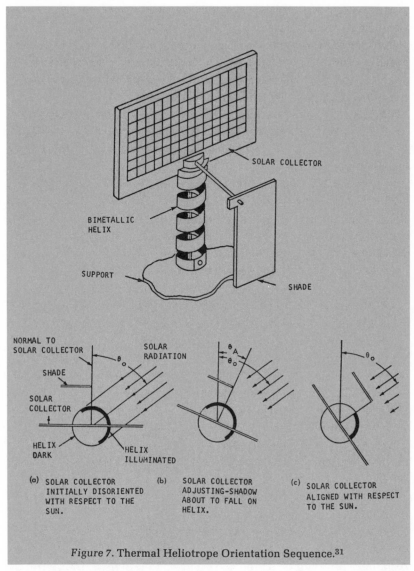

Figure 7. Thermal Heliotrope Orientation Sequence.[31]

area of the helix illuminated by the sun's rays is thereby increased causing the shade to rotate in the clockwise direction until the energy balance on the helix is restored.[31]

The bimetal considered the prime candidate for terrestrial use has a high expansion component of 72% Mn-18% Cu-10% Ni and a low expansion component of 36% Ni-64%Fe, commonly referred to as Invar. This bimetal is one of the most thermally active and one of the least expensive. The thermal heliotrope is a promising passive orientation device which probably could be produced in large quantities at low unit cost, and thus reduce the cost of tracking the sun for the collection of solar energy.

Instead of steering a single concentrator Gunter[32] proposed a faceted solar concentrator in which the separate flat reflecting facets were rotated by a single mechanism. Each facet is rotated at exactly the same speed to keep the reflected sunlight focused on a fixed heat collecting element. Another approach is to focus many separate flat mirrors onto a single point. The difficulty with this latter system is that each mirror requires a separate steering mechanism; but if large numbers are used, they may lend themselves to the economics of mass production.

The second approach to reducing the concentrator cost is to fix the reflector and move the heat collecting element. The problem with this is that the standard reflecting surfaces are only in focus for one of the sun's directions. The parabolic cylinder and parabolic concentrators are only in focus when the sunlight is incident along the axis of the parabola, so the problem with such fixed collectors, as proposed by Steward,[33] is that the focus is severely degraded whenever the incident direction of the sunlight is significantly off axis. This results in a reduced concentration factor and reduced collection temperature.

Recently a new type of reflecting surface was proposed[34] that remains in focus for any incident sun angle. It is composed of long, narrow, flat reflecting elements arranged on a concave cylindrical surface. The angles of the reflecting elements are fixed so that the focal distance is twice the radius of the cylindrical surface. The focus is always sharp for parallel light of any incident direction. The point of focus lies on the reference cylindrical surface, so the heat exchanger pipe can be supported on arms that pivot at the center of the reference cylinder. This greatly simplifies the positioning of the heat exchanger.

Chapter 4

HEATING FOR HOUSES AND BUILDINGS

The Committee on Science and Astronautics of the U.S. House of Representatives has concluded that "the most promising area for the application of solar energy within the next 10–15 years, on a scale sufficient to yield measurable relief from the increasing demands upon fossil fuels and other conventional energy sources, is the use of solar energy for space heating, air conditioning, and water heating in buildings".[35] As is seen from Table 3, energy for space heating, air conditioning, and water heating in building services accounts for about 25% of the total energy consumption in the United States, and is presently supplied almost totally by the

Table 3. Energy Consumption in the United States by End Use, 1960–68
(Trillions of BTU and Percent/Year)

Sector and End Use	Consumption 1960	Consumption 1968	Annual Rate of Growth (%)	Percent of National Total 1960	Percent of National Total 1968
Residential:					
Space heating	4,848	6,675	4.1	11.3	11.0
Water heating	1,159	1,736	5.2	2.7	2.9
Cooking	556	637	1.7	1.3	1.1
Clothes drying	93	208	10.6	.2	.3
Refrigeration	369	692	8.2	.9	1.1
Air conditioning	134	427	15.6	.3	.7
Other	809	1,241	5.5	1.9	2.1
Total	7,968	11,616	4.8	18.6	19.2

Table 3—*Continued.*

Sector and End Use	Consumption 1960	Consumption 1968	Annual Rate of Growth (%)	Percent of National Total 1960	Percent of National Total 1968
Commercial:					
Space heating	3,111	4,182	3.8	7.2	6.9
Water heating	544	653	2.3	1.3	1.1
Cooking	98	139	4.5	.2	.2
Refrigeration	534	670	2.9	1.2	1.1
Air conditioning	576	1,113	8.6	1.3	1.8
Feedstock	734	984	3.7	1.7	1.6
Other	145	1,025	28.0	.3	1.7
Total	5,742	8,766	5.4	13.2	14.4
Industrial:					
Process Steam	7,646	10,132	3.6	17.8	16.7
Electric drive	3,170	4,794	5.3	7.4	7.9
Electrolytic processes	486	705	4.8	1.1	1.2
Direct heat	5,550	6,929	2.8	12.9	11.5
Feedstock	1,370	2,202	6.1	3.2	3.6
Other	118	198	6.7	.3	.3
Total	18,340	24,960	3.9	42.7	41.2
Transportation:					
Fuel	10,873	15,038	4.1	25.2	24.9
Raw materials	141	146	.4	.3	.3
Total	11,014	15,184	4.1	25.5	25.2
National Total	43,064	60,526	4.3	100.0	100.0

Note: Electric Utility consumption has been allocated to each end use.
Source: Patterns of Energy Consumption in the United States.[14]

combustion of high quality fossil fuels. The sources which supply this energy are depicted by Figure 8. Space heating accounts for more than half of the total residential energy consumption. Space heating alone for homes and businesses accounts for 18% of all energy consumption in the United States. In the South, where solar energy is most available, practically all residential energy comes from gas or electricity, and even in the South about half this energy is used for space heating (Figure 9). Space heating and water heating account for over ⅔ of all residential energy consumption in the South.

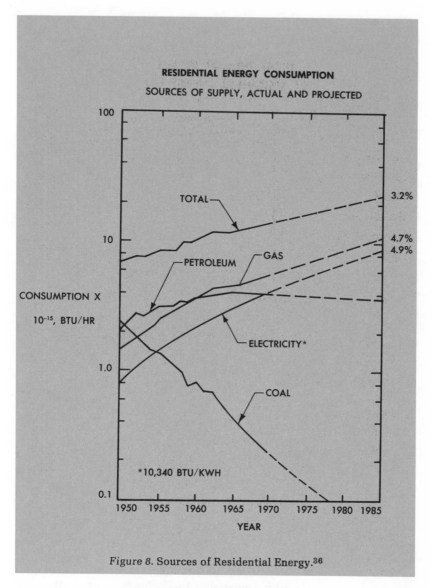

Figure 8. Sources of Residential Energy.[36]

Flat Plate Collector Systems

A typical solar heating system employing a flat plate collector is illustrated by Figure 10. A flat plate collector located on a south-ward sloping roof heats water which circulates through a coil in the

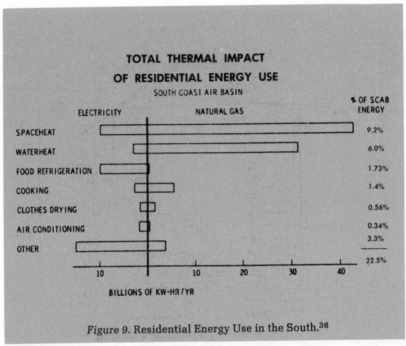

Figure 9. Residential Energy Use in the South.[36]

hot water tank, then through a coil in a large warm water tank before being returned to the collector. In most areas of the country the heat transfer fluid flowing through the collector should be an anti-freeze solution to prevent freezing of the fluid in the collector tubes in the winter. The system shown in Figure 10 provides for two levels of heat storage; the hottest water which is stored in the hot water tank is used for building services, and the warm water in the large tank heats water circulating through pipes in the house. The heat reservoir for a single dwelling could be a tank 10 ft in diameter, four ft deep, and insulated on all sides. An auxiliary heating system is necessary to provide heat during extended cold cloudy periods when the supply of solar heat is not adequate.

One study compared the cost of solar heating with gas, oil and electric heating costs by amortizing the solar system capital cost over 20 years at 6% interest.[37] Solar heating costs were calculated for present $4/ft^2 flat plate collectors and for anticipated near-term collector costs of $2/ft^2. The results of these calculations for eight U.S. cities are given in Table 4. Since these data were compiled,

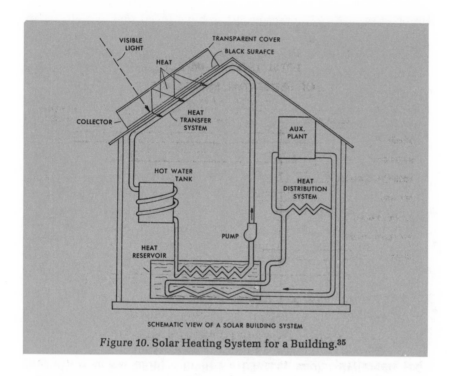

SCHEMATIC VIEW OF A SOLAR BUILDING SYSTEM

Figure 10. Solar Heating System for a Building.[35]

Table 4. Costs of Space Heating in 1970 Dollars/MBTU[37]

Location	Optimized Solar Heating Cost in 25,000 BTU/degree-day House, Capital Charges @ 6%, 20 years		Electric Heating, Usage 30,000 kwh/year	Fuel Heating Fuel Cost Only	
	Collector @ $2/ft²	Collector @ $4/ft²		Gas	Oil
Santa Maria	1.10	1.59	4.28	1.52	1.91
Albuquerque	1.60	2.32	4.63	0.95	2.44
Phoenix	2.05	3.09	5.07	0.85	1.89
Omaha	2.45	2.98	3.25[3]	1.12	1.56
Boston	2.50	3.02	5.25	1.85	2.08
Charleston	2.55	3.56	4.22	1.03	1.83
Seattle-Tacoma	2.60	3.82	2.29[2,3]	1.96	2.36
Miami	4.05	4.64	4.87	3.01	2.04

Notes: [1] Electric power costs are for Santa Barbara; electric power data for Santa Maria were not available.

[2] Electric power costs are for Seattle.

[3] Publicly owned utility.

interest rates have increased but so have fossil fuel and electricity prices, so the general conclusions are still valid. The two major accomplishments of this study were optimization of the design for a solar heating system and its major components, and establishment of realistic costs of solar heating compared with conventional heating under a variety of conditions, using methods which can be applied to buildings of any size and construction in any location where adequate weather data are available. The collector size for minimum solar heat cost for a 25,000 BTU/degree-day (BTU/DD) house in six locations was found to range from 208 ft² (Charleston, S.C.) to 521 ft² (Omaha, Nebraska), corresponding to 55% of the respective annual heating loads. Degree-days are calculated by multiplying the number of degrees the ambient average temperature is less than 65°F by the number of days each temperature less than 65°F is achieved. In Santa Maria, California a 261 ft² collector can supply 75% of the annual heat requirement. In most situations the cost of solar heat near optimum levels is rather insensitive to collector size and the corresponding fraction of load carried. Costs rise sharply, however, if designs are based on carrying large fractions (over 90%) of the load. In structures having smaller or larger heat demands than 25,000 BTU/DD, optimum collector size is approximately proportional to the demand parameter. Heat storage capacity for minimum solar heating cost in nearly all practical situations is 10–15 lb of water (or its thermal equivalent) per square foot of collector, which is the equivalent to one to two days average winter heating requirement. Two glass covers (double glassing) for the flat plate collectors yield minimum solar heating costs in most locations. One cover is optimal in the warmer climates represented by Phoenix and Miami. Heating costs are the same for one or two covers in climates such as Albuquerque and Santa Maria. The collector slope which yields the minimum solar heat cost is 10–20° greater than the latitude, but there is only a slight variation in cost over a range of inclinations between the latitude angle and 30° higher than the latitude.

The cost of solar heat in systems of optimum design was usually in the range of two to three dollars per million BTU, and substantially below the cost of electric heat in six of the eight locations examined. In comparison with gas and oil heating, solar heating alone was more expensive in six of the eight locations analyzed in 1970. But in Santa Maria and Albuquerque there were combinations of solar and fuel systems with total costs equal to or below those of corresponding conventional heating. In six of the eight cities there were optimum (minimum cost) combinations of solar and electric heating, the best mix being obtainable by determining marginal costs of increasing the solar heat proportion. The portion of total load supplied by solar heat under these conditions generally lay between 60 and 90 percent. Rising costs of heating with oil and gas have now approached or surpassed solar heating costs in U.S. areas of large population. It is also probable that solar heating costs will decrease as solar systems become commercially available. Conditions conducive to economical solar heating are moderate to severe heating requirements, abundant sunshine, and reasonably uniform heat demand during the period when heat is needed.

Materials costs have been estimated for a flat plate collector using a single glass cover, water as the coolant, and polyurethane foam insulation, and a collector efficiency of greater than 50% at outlet temperatures up to 200°F. On a production basis cost was determined to be between $1.15/ft² and $1.90/ft².[29] However, February 1974 prices of suitable heat exchangers were in excess of $2.00/ft² and selective coating prices are uncertain, because large quantities of the components were not yet being produced on a large volume production basis.

Table 5. Materials Cost for a Flat Plate Collector[29]

Component	Material	Cost $/Ft²
Substrate/heat exchanger	Aluminum or steel	0.60 to 0.90
Cover plate	Glass	0.25 to 0.30
Thermal insulation	Polyurethene	0.25 to 0.35
Selective coatings	Oxides, coatings	0.05 to 0.35
Total		1.15 to 1.90

Several studies have been conducted to determine optimal control systems for solar home heating systems, such as the one illustrated by Figure 11. The main object of the control system is to extract heat from the solar collector when it is available, but to shut off the flow through the collector whenever the collector tem-

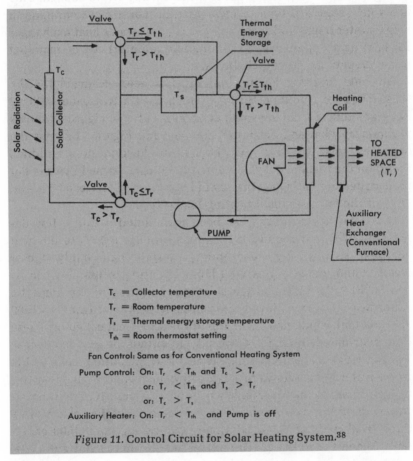

Figure 11. Control Circuit for Solar Heating System.[38]

perature drops below the storage temperature. In this system a separate auxiliary heater is provided. The pump circulates water through the collector whenever the collector outlet temperature exceeds the storage temperature. If the room temperature is lower than both the collector temperature and the thermostat setting, water from the collector is circulated directly through the heating pipes in the house. If the room temperature is lower than both the thermostat setting and the storage temperature, but higher than the collector temperature (such as at night), hot water from the storage tank is circulated through the room. Thus, the solar heat is transferred directly to the room if it is too cool, and transferred to the storage tank for later use if the room is already warm enough. This is a fairly standard type of solar thermal control system using a single water pump and three valves. The hot water heat exchanger for heating air (similar to an automobile radiator) can be installed in a conventional forced air furnace.

An 8000 ft² solar heated building has been designed for the Massachusetts Audubon Society which uses a two-pane 3500 ft² flat plate collector facing south at an angle of 45°. Figure 12 shows preliminary plan and elevation sketches and Figure 13 shows the proposed solar building and the current headquarters building. Based on the results of Tybout and Löf[37] it is estimated that the flat plate collector heating system should account for between 65% and 75% of the total seasonal heating load.

One solar heated house has been maintained within a few degrees of 70°F year round for 13 years, with up to 95% of the heat per year supplied by solar energy, despite half-cloudy winter weather and temperatures well below freezing (often between 0° and 32°F).[40, 41] Additionally a substantial portion of the domestic water heating was achieved by solar heating. Water from the 1600 gal steel tank is pumped to the top of the solar heat collector. There it is distributed in small streams to hundreds of grooves on the black corrugated heat absorber. A gutter at the bottom collects the warm water, which flows to a 275 gallon domestic water preheater tank and then to the main tank, as shown in Figure 10. The warmed water, in addition to pre-heating the domestic water, also warms three truckloads of fist-sized stones around the main 1600 gallon tank. When the living quarters need heat, a thermostat automatically starts a ¼ horsepower blower which blows air through the warmed stones around the warm tank of water, thus warming the air which flows into the living quarters. When the house is warmed sufficiently, the thermostat automatically stops the blower. This system has maintained the home temperature at 70°F, plus or minus 2°F, for four cloudy days in mid-December. During the hot summer, water is pumped at night up to the north-sloping roof section, cooling the water in the tank and surrounding stones. During the day the blower circulates air through the cooled stones to cool the house.

Flat plate collectors are also used for directly heating air for

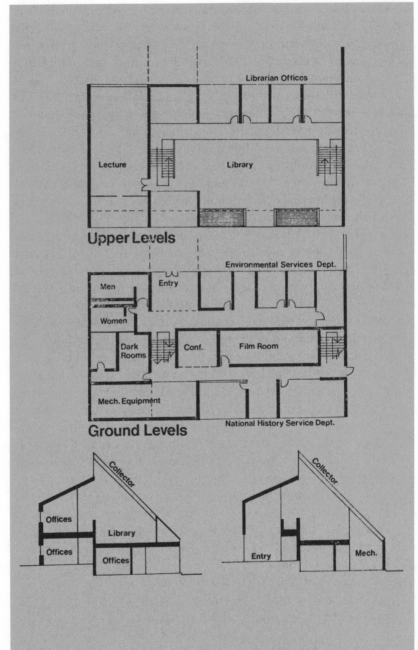

Figure 12. Plan and Elevation Sketches of Proposed Solar Buildings.[39]

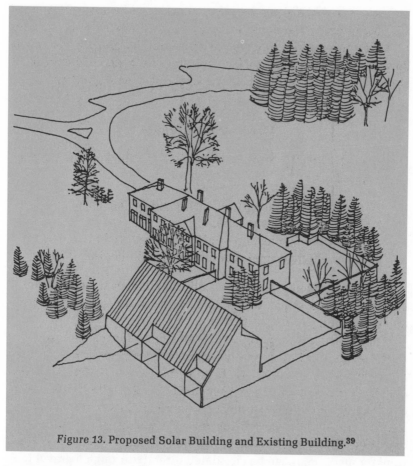

Figure 13. Proposed Solar Building and Existing Building.[39]

house heating. Water is usually used because of the simple storage system, which is just an insulated tank. The simplest air heater is a flat black plate covered by a transparent sheet, with air flowing in the gap between. However, higher temperatures are achieved if the air flows through or beneath the black absorbing surface, and the air gap beneath the transparent cover and plate is stagnant. A good collecting surface is a V-corrugated absorber plate with a spectrally selective coating (absorptivity 0.80 in the visible, 0.05 in the infrared). Absorbers of this type heated air to 170°F with 40% collection efficiency for an insolation of 160 BTU/ft²/hr and an ambient dry bulb temperature of 74.6°F. For an insolation of 300 BTU/ft²/hr a temperature of 210°F was reached with 40% collection efficiency.[42] The maximum temperature of the air can be increased an additional 10–15°F with no loss in collector efficiency by allowing the air to flow over the absorbing surface and then back under the absorber (2 passes) instead of the standard single-pass configuration.[43]

Concentrator Systems

Concentrators offer several advantages for the heating and cooling of buildings:

1. Higher collection efficiencies result in smaller collectors
2. More compact heat storage
3. Year round collection of high temperature heat
4. More efficient operation of absorption cooling devices

Also, higher temperature heat collection makes the generation of electric power possible, with waste heat used for space heating and air conditioning.

When concentrators are used, water is no longer an acceptable heat transfer fluid so air or a commercial heat transfer fluid is used. Steward[33] proposed a 962 ft² fixed cylindrical reflector with movable heat exchanger to collect heat at 500°F. Russell's[34] fixed mirror concentrator has the advantage of remaining in sharp focus for all incident sun angles, permitting the efficient collection of heat at 500°F or more during most of the day. If air is used as the heat transfer medium, it can be circulated directly through a gravel tank for heat storage, then brought from the gravel tank to the house, as required, for heating and other applications. More compact heat storage is possible with phase-change materials such as Glauber's salt (sodium sulfate decahydrate, Na_2SO_2-$10H_2O$). Water, rocks and a typical phase-change material may be compared as follows (Table 6).[44] Water can store heat over a range of temperatures approaching 200°F, and rocks can store heat (or coolness) at any conceivable temperature, but phase change materials melt and solidify at one temperature. Thus rocks and water can store heat in the winter and "coolness" in the summer, whereas two separate salt systems would be required to accomplish this. Phase change ma-

Table 6. Thermal Storage of One Million BTU with 20°F Temperature Change

	Water	Rocks	Phase Change Material
Specific Heat (BTU/lb°F)	1.0	0.2	0.5
Heat of Fusion (BTU/lb)	—	—	100
Density (lb/ft³)	62	140	100
Weight (lb)	50,000	250,000	10,000
Volume (ft³) with 25% passage	1,000	2,150	125

terials also cost more per BTU of heat storage than water or rocks; gravel costs are about $5.00/ton. The advantages of the phase change material are, of course, considerably reduced weight and volume.

Roof Ponds

Perhaps the simplest technique for heating and cooling a house is to locate a pond of water 6–10 inches deep on the roof. The pond is covered by thermally insulating panels which can be open or closed. In the winter all the water is enclosed in polyethylene bags atop a black plastic liner. Sunlight heats the water to about 85°F during the day. At night, the insulating panels are lowered to prevent loss of the heat to space. During the summer, the insulating panels are open at night and closed during the day, so the water is cooled by radiation to space at night. Tests with a small 10 ft by 12 ft structure in Phoenix, Arizona, showed that temperatures were maintained close to 70°F year round by pulling a rope twice a day, even though ambient temperatures ranged from subfreezing to 115°F.[45] A new two bedroom house with a 10 inch roof pond is now being tested (Figure 14) in California at Atascadero near Paso Robles which has recorded temperature extremes of 10°F and 117°F.[45] The horizontal roof collector is not expected to meet the full heat demand because ambient air temperatures are lower, cloud cover is greater, and the location is two degrees more northerly than the Phoenix location of the test room. Summer cooling, however, should be better than at the Phoenix location. The roof pond is not visible at ground level. The house is to be occupied for one year while it is evaluated by professors from the California Polytechnic

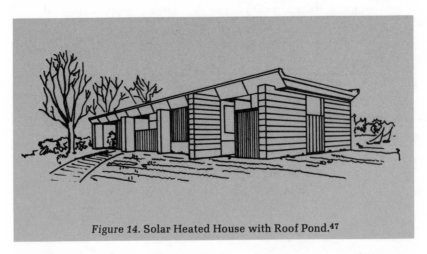

Figure 14. Solar Heated House with Roof Pond.[47]

University with financial support from the Department of Housing and Urban Development.[47]

The French System

An economically attractive natural circulation solar heating system has been developed in France and tested with several full-sized houses. A south facing concrete wall is painted black and covered with glass, with an air gap between the wall and the glass. Sunlight heats the concrete wall and air circulates through vents from the room, rises as it is heated in the air gap, and flows out into the room through vents near the ceiling. The concrete wall acts both as heat collector and heat storage medium, since the warmed wall continues to provide heat in the evening. During the summer the room vent near the ceiling is closed and an exterior vent opened, so the system operates as before but the warmed air flows outside, drawing cooler air into the room from the north side of the house. This type of solar heating system can supply half or more of the year-round heating requirements of a house in France.

Costs

The present range of costs for solar heat is from less than $2.00 per million BTU (MBTU) to about $5.00 per MBTU, depending on

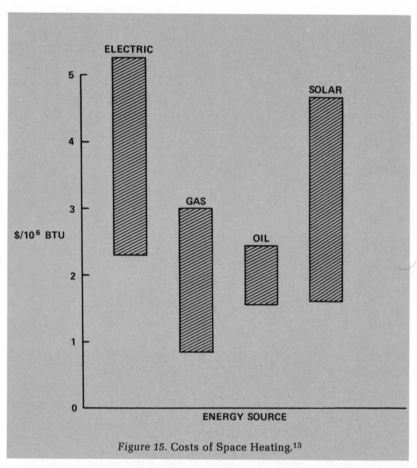

Figure 15. Costs of Space Heating.[13]

climate, the type of system used, and prevailing materials and labor costs. As seen from Figure 15, solar heating costs tend to compare favorably with electric heat, but unfavorably with natural gas. As prices of oil and gas rise and the technology for solar heating advances, solar heating systems will become increasingly competitive.

Chapter 5

SOLAR WATER HEATERS

Solar water heaters are currently in widespread use throughout many sunny areas of the world. A common arrangement is to have a flat plate solar collector on the roof that provides hot water by natural circulation to a tank located higher on the roof. The roof tank can be designed to look like a chimney or located in the attic. In Japan there are about 2½ million solar water heaters of several different types currently in use.[48] The Japanese units employ a storage tank and collector as an integral unit, whereas in other countries the storage tank is usually separated. The simplest and oldest type is a flat open tank on the roof (about $10 with a black bottom) which supplies water at 130°F in the summer and as high as 80°F in the winter. Since the water is sometimes contaminated by dust, a polyethylene film covering the tank can be added for a few extra dollars. The transparent cover lasts about three years, and increases the water temperature as well as preventing contamination. The standard heater size is about 3 ft wide, 6 ft long and 5 inches deep. These flat tank water heaters are cheap, but suffer a major disadvantage because they must be mounted horizontally, so they are not very effective in the winter when the sun is low. Closed pipe collectors can be mounted at a more optimum angle to the sun and thus provide hotter water during the winter months. The pipes are made of glass, plastic or stainless steel painted black and mounted in a frame covered with glass or transparent polyethylene plastic. The cost of these units range from $100–$200. The purchasing of solar water heaters has declined since 1967 because of the availability of convenient and inexpensive heaters using fuels such as propane gas, however the recent rapid escalation of fuel prices has increased solar water heater sales again.

In the United States about 60–70 ft^2 of collector can supply 75% of the water heating needs of apartments. One study now underway is Project SAGE (Solar Assisted Gas Energy)[36] in southern California, a joint project of the Jet Propulsion Laboratory and the California Gas Company, which is studying the technical and economic aspects of a solar assisted gas and electric water heating system for a typical Southern California apartment building. Figure 16 illustrates a solar-electric hot water system for an apartment complex, with a single collector and storage tank, which reduces

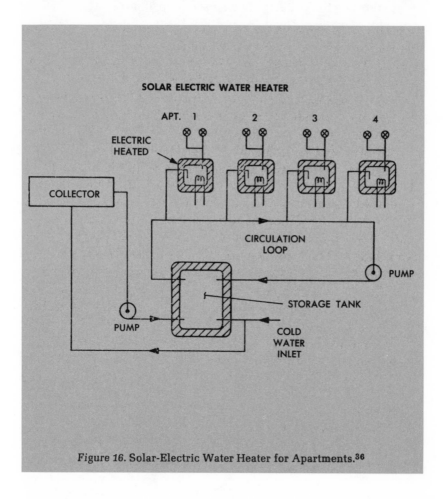

Figure 16. Solar-Electric Water Heater for Apartments.[36]

the cost of collecting the solar heat for the apartment. The cost of
the solar collection and storage is thus part of the cost of building
and maintaining the apartment building, and is included in the rent.
The electric power consumed, however, is paid for by the individual
user in his electric bill. This aspect of the system is attractive from
the viewpoint of the apartment owner since it provides account-
ability for the consumption of hot water during periods when the
solar input alone is not adequate. The same general type of solar
collector can be used to preheat water before it enters a conventional gas water heater. Water heating in a freezing climate requires
that an intermediate heat transfer fluid (antifreeze solution) circulate through the collector in a closed loop and transfer its heat to
water in a heat exchanger (Figure 17). If the collector temperature
is higher than the cold water inlet temperature, which is usually
the case when the sun is shining on the collector, the pump is turned
on and fluid from the collector circulates through a coil in the
storage tank, thus preheating the water in the tank before it enters
the conventional heaters. Solar heat is thereby used year round to
reduce the consumption of gas or electricity for heating water.
During parts of the summer all of the heat can be supplied by the
solar collector. For the closed loop system the water flowing
through the collector can be maintained at a lower pressure than
the water in the storage tank.

The costs of electric, gas, and solar-assisted gas water heating
are compared in Figure 18. It is clear that as gas prices rise, and as
solar collector costs decrease from $4/ft², solar-assisted gas water
heaters will become cheaper than gas heaters alone. Already, solar
assisted electric heating is cheaper than electric water heating
alone, because of its high cost. At the present time gas is not being
supplied to new units, because of short supply, in some areas of the
country. The cost comparisons shown in Figure 18 are based on a
discount rate of 10%/year, a system life of 10 years, and an apartment size of 50 units.[36]

The design and performance of a large forced-circulation water
heater of the same general configuration as that considered in the
SAGE study has been evaluated.[50] The flat plate collector consisted
of a 28 gauge blackened aluminum sheet attached to 1.9 cm

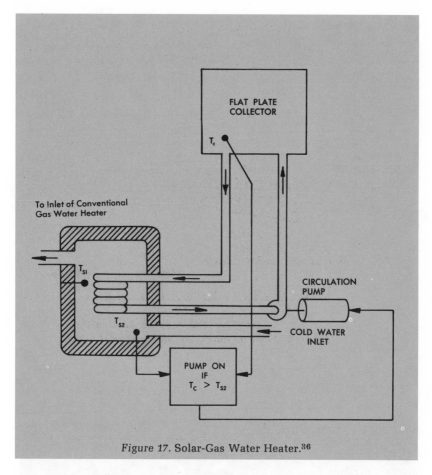

Figure 17. Solar-Gas Water Heater.[36]

diameter galvanized pipe with 10 cm spacing, as shown in Figure 19. This collector configuration is optimized for maximum heat collection per unit cost.[51-53] A single 3 mm glass sheet covers the collector plate. The vertical, cylindrical storage tank had a height of twice the diameter to reduce the heat loss when the hottest water is located in the upper part of the tank. The total collector area was about 100 ft²; it heated the water in the tank to as hot as 130°F with a collection efficency of 50%. The pump consumed only 7 kilowatt hours per month. Detailed computer studies of these types of solar water heaters have been performed.[54]

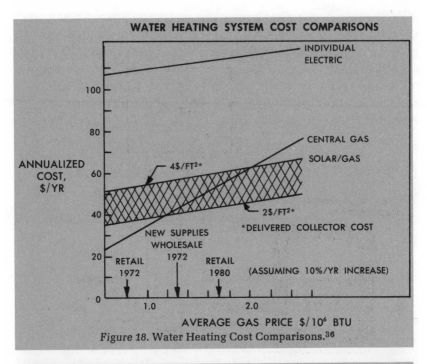

Figure 18. Water Heating Cost Comparisons.[36]

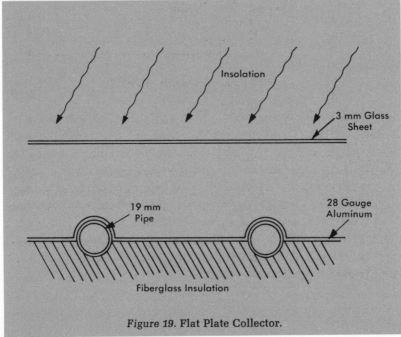

Figure 19. Flat Plate Collector.

Solar water heating has been quite popular in Israel, and by 1965 over 100,000 units had been installed.[55] One reason is that until recently the cost of electricity and heating fuels has been high enough to make solar heating more economical. The first solar water heaters in Israel were sold with a three-year guarantee. This was soon raised to 5 years, and for a small additional cost could be extended to 8 years.

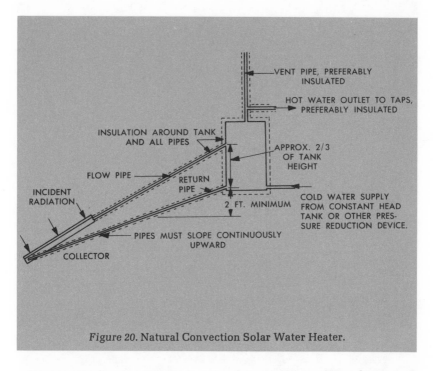

Figure 20. Natural Convection Solar Water Heater.

Figure 20 shows a typical natural convection water heater. A small reverse flow occurs when the collector is cooler than the water in the tank, which has a certain advantage in the winter because it prevents water from freezing in the collector. One series of tests measured the overall performance as affected by seasonal variations, type of transparent covering, insulation, height of storage tank and location of the point joining the flow pipe to the storage tank. The average efficiency was about 50% with a polyvinyl fluoride collector covering, and about 55% with glass. There was little effect on the efficiency of changes in insulation and seasonal variations.[56]

In the United States solar water heating should probably be utilized in all new apartment and housing units in areas with mild winters. In more northern climates where winter temperatures drop well below freezing, natural circulation systems such as shown in Figure 20 and closed-loop systems as shown in Figure 17 can be used.

Chapter 6

AIR CONDITIONING

Solar cooling is usually accomplished by using solar heat to operate a thermal absorption type refrigeration system. In regular electrically operated refrigerators, a vapor such as ammonia is condensed to a liquid with a motor-driven pump, and the heat released is removed by blowing air through the condenser. The liquid is then vaporized in an insulated box and heat is absorbed by the vaporization process to give the cooling effect. In solar refrigeration, the cycle is similar except that the pressure is produced by heating a concentrated solution of ammonia to give a high vapor pressure, instead of compressing the vapor mechanically. There are two connecting, gastight vessels, one of which contains liquid ammonia and the other a very concentrated solution of salt in liquid ammonia. The salt solution has a much lower vapor pressure. The liquid ammonia vaporizes in its compartment, thereby cooling it, and dissolves in the salt solution contained in the other compartment. The system is regenerated by using focused solar radiation to raise the temperature of the salt solution so the vapor pressure of ammonia in the solution exceeds the vapor pressure of the pure liquid ammonia in the second compartment. In this way, the operating cycle produces cooling by evaporating ammonia as it goes into the concentrated solution of salt, making it more dilute; and the solar regeneration drives out the ammonia from the diluted salt solution to produce pure ammonia and leaves a more concentrated solution. Another cycle uses a concentrated solution of lithium bromide to absorb water vapor which causes liquid water in another compartment to vaporize to provide cooling. The lithium bromide solution is concentrated again by heating the dilute solution with solar heat, and the system is operated continuously.[57]

A continuously operating absorption air-conditioning system was built and tested in the early 1960's at the University of Florida.[58] Hot water was used to heat a high-concentration, ammonia-water solution (50–60% ammonia by weight) in a generator, driving the ammonia out of the solution. The ammonia vapor was then condensed and expanded through an adjustable expansion valve, entering the evaporator as a two-phase mixture. The liquid component evaporated, cooling the water circulating through tubes in the evaporator, and then was reabsorbed into the water, and the ammonia solution was pumped back to the generator to repeat the cycle. Ten 4 ft by 10 ft flat-plate solar collectors provided the hot water to operate the air conditioner. The absorbing surfaces were tubed copper sheets painted flat black, placed in galvanized sheet-metal boxes with two inches of foam-glass insulation behind, and a single glass cover. The system was operated with heating water temperatures ranging from 140–212°F. The maximum cooling effect was 3.7 tons, and steady operation was achieved with 2.4 tons of cooling.

Solar powered air conditioning systems can also be driven by an organic Rankine cycle engine. Solar heat could be used to vaporize an organic fluid at a temperature between 160°F and 280°F to drive a Rankine cycle engine, which in turn would drive the compressor of a vapor-compression air conditioning system.[59] The coefficient of performance should compare favorably with absorption air conditioning systems, but at the present time none have been built. Except for the northern-most part of the country, combined solar heating and cooling is cheaper than solar heating or cooling alone. Solar costs were calculated based on $2/ft² collectors, amortization over a 20 year life, 8% interest, 1970 prices, and an additional $1,000 capital cost for solar air conditioning over electric, gas or oil air conditioning.[60] Water heating was included, and the water storage cost was taken to be $0.05/lb water. As seen from Table 7, solar heating is very costly in Miami where not much heat is needed, and likewise solar air conditioning is not economical in northern climates that require little air conditioning. The primary reason that the combined system is usually cheaper than either alone is that it

Table 7. Heating and Cooling Costs

| | | $ Per Million BTU | | |
	Oil or Gas	Electric	Solar Heating	Solar Cooling	Solar Combined
Albuquerque	0.95	4.63	2.01	3.24	1.70
Miami	2.04	4.87	11.63	2.19	2.07
Charleston	1.03	4.22	3.34	3.50	2.47
Phoenix	0.85	5.07	2.86	2.05	1.71
Omaha	1.12	3.25	2.93	5.41	2.48
Boston	1.85	5.25	3.02	8.74	3.07
Santa Maria	1.52	4.28	1.57	14.60	2.45
Seattle	1.96	2.29	3.15	19.63	3.79

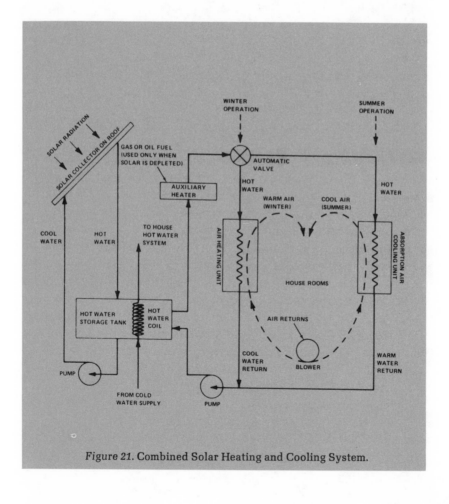

Figure 21. Combined Solar Heating and Cooling System.

permits both summer and winter utilization of the solar collector—
the most expensive part of the system. Thus, in most parts of the
U.S., the economics favor combined solar heating and cooling,
rather than either alone. Figure 21 illustrates such a combined system using a common collector, storage tank, auxiliary heater, and
blower for both heating and air conditioning. It should be kept in
mind that oil and gas prices have risen substantially since this study
was performed.

Chapter 7

ELECTRIC POWER GENERATION

A variety of approaches have been used for converting solar energy into electricity, including solar-thermal conversion, photovoltaic devices, and bio-conversion. Sunlight is an abundant, clean source of power; all that is required is the development of technology to economically convert this energy into electricity.

The NSF/NASA Solar Energy Panel[13] identified the various possible steps leading from solar radiation to power delivered to the consumer (Figure 22). In this scheme plants, rivers, winds, ocean currents and ocean temperature gradients are considered natural collectors of solar energy. Solar energy can also be collected directly as heat, or converted into electricity via the photoelectric effect. If collected as heat, it can be stored for use when the sun is not shining. The heat can be used to operate a power plant or to produce a chemical fuel, such as through the thermochemical production of hydrogen. The fuel can be stored and used as needed to produce electric power, such as with a hydrogen-air fuel cell.

With so many possible approaches available for the production of electric power, the problem then is to choose the approach which is most cost-effective for a specific application. This is sometimes difficult since technology is advancing rapidly in most of these areas, and the comparative economics becomes uncertain. At present, the two most promising technological approaches are photovoltaic conversion with electrical storage, and solar-thermal conversion with heat storage for night-time operation.

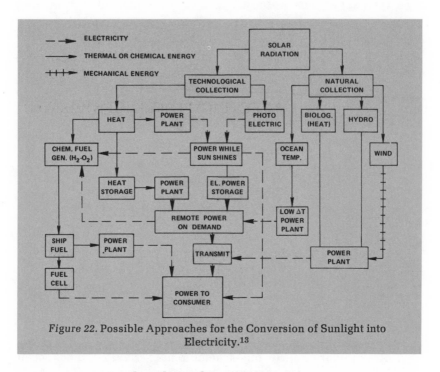

Figure 22. Possible Approaches for the Conversion of Sunlight into Electricity.[13]

Solar-Thermal Power Generation

The two main approaches to solar-thermal power generation are the solar furnace approach, in which sunlight reflected from many different locations is concentrated on a single heat exchanger, and the solar farm, with large numbers of linear reflectors focusing solar radiation on long pipes which collect the heat.

The tower concept proposed in 1949 is a good example of the solar furnace approach.[61] A large number of flat mirrors covering a large area of land independently focus sunlight onto a boiler, which is mounted at the top of a tower located near the center of the field of mirrors to produce high temperature steam for driving a turbine. A 50 kilowatt plant has been built and operated in Italy.[62] An advantage of this system is that the separate mirrors and steering mechanisms can be mass produced, and the smaller reflectors are less subject to high wind loadings than a single large steerable concentrator of the same total collector area.

A system has been proposed with over a thousand 10 ft mirrors covering a 6000 ft diameter circle of about one square mile area to reflect sunlight onto the boiler on top of a 1500 foot high tower (Figure 23). Each mirror would be steered separately by a heliostat

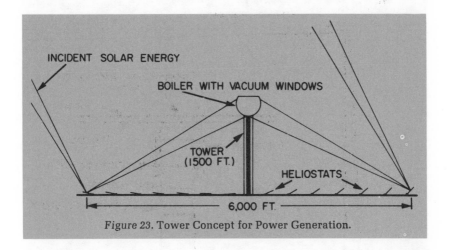

Figure 23. Tower Concept for Power Generation.

as shown in Figure 24. Since the major expense of solar energy collection employing a solar furnace would be the heliostats, considerable research needs to be done in order to develop a heliostat which could be economically mass produced. The 150 ft diameter, 1500 ft high tower would cost about $15 million. The boiler could be made of steel and operate in the 1000°C range, and the solar image size at the boiler would be 31 ft in diameter. The outer boiler surface would be black and could be surrounded by an evacuated glass envelope. About 20% of the incident solar energy would be lost upon reflection by the mirrors, and another 6% lost by reflection from the boiler glass envelope. If 45% of the land area is covered with mirrors the boiler could collect 630 BTU/day per sq ft of mirrors in the Southwest U.S. in the winter, 1320 BTU/ft² day in the spring and fall, and 1620 BTU/ft² day in the summer. The total cost of heat collected by this plant is estimated at $0.48 per MBTU,[63] which is competitive with the cost of fossil fuels delivered in large quantities to a power plant. This cost estimate is based on a $2/ft² cost for the mirrors and heliostats and $15 million for the tower.

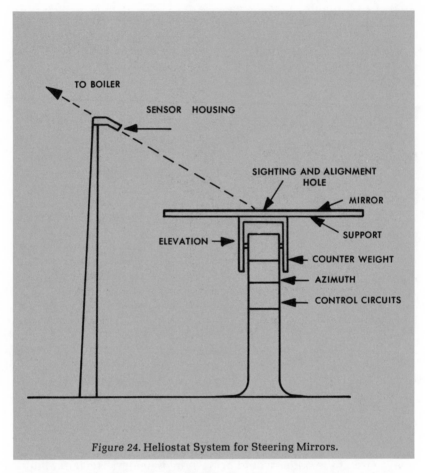

Figure 24. Heliostat System for Steering Mirrors.

The heliostats must automatically aim the mirror with an accuracy of 0.2° in the presence of winds.

A megawatt solar furnace has been developed in France employing heliostats with 20 inch square flat glass mirrors and a fixed parabolic concentrator on the side of a nine-story building.[64] The flat mirrors reflect sunlight toward the fixed parabolic concentrator that focuses the sunlight. The heliostats and mirrors cost $21/ft². Major problem areas that need to be investigated are heat shock from the many thermal cycles which result from clouds passing in front of the sun and the absorption–reflection–radiation characteristics of potential boiler surfaces operating at high temperatures and high heat fluxes.

Another similar power plant system uses arrays of heliostat-guided mirrors to focus sunlight into a cavity-type boiler near the ground to produce steam for a steam turbine electric power plant. Sunlight striking the mirrored faces of the heliostat modules is reflected and concentrated in the cavity of the heat exchanger. To demonstrate the feasibility of generating large amounts of electric power with this system, Martin Marietta and the Georgia Institute of Technology, with support from the National Science Foundation, are preparing to generate about 300 kilowatts of electric power (KWe) using the French furnace in Odellio.[65] One useful application for this system would be to augment existing hydroelectric plants whose generating capacity exceed their water supply. With solar augmentation the power production for the central six hours of daylight is taken over by the solar system, so hydroelectric generation during the remaining hours is augmented 33% by the water saved during solar operation. The solar plant does not need to be anywhere near the hydroelectric facilities, as long as they are connected to the same power grid. This solar plant can also be used for peaking in areas where peak power demand is during daylight hours.

Solar farms have been proposed using parabolic trough concentrators or other types of concentrators to focus sunlight onto a central pipe surrounded by an evacuated quartz envelope. Heat collected by a fluid flowing through the pipes could be stored at temperatures over 1000°F in a molten eutectic and used as required to produce high enthalpy steam for electric power generation.[66] Another approach is to store the heat in rocks, and extract the heat as required to generate steam on demand, as illustrated by Figure 25.

For large scale solar-thermal electric power generation to become economically feasible, the cost of the collector must not exceed about one dollar per square foot. However, concentrating solar collectors that must be steered to follow the sun cost more than $4/ft², and a major part of this cost is the steering mechanism and the mechanical structure which must withstand reasonable wind loadings. The fixed mirror concentrator, on the other hand, does not

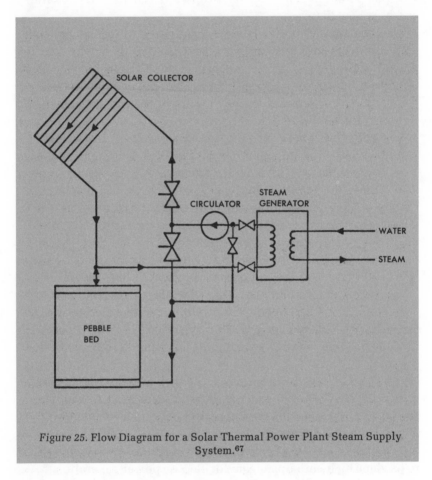

Figure 25. Flow Diagram for a Solar Thermal Power Plant Steam Supply System.[67]

have to be steered and need not be self-supporting, so fabrication of these concentrators should be cheaper than for steerable reflectors.[67, 68] Since the point of focus always lies on the reference cylindrical surface, the heat exchanger pipe can be supported on arms that pivot at the center of the reference cylinder. This greatly simplifies the positioning of the heat exchanger.

A proposed power plant for the Southern California desert would be arranged in modules (Figure 26) with 30 ft wide fixed mirror concentrators arranged in a 1500 by 1880 ft array, and a gravel tank for heat storage and the steam generator located in the center. Air

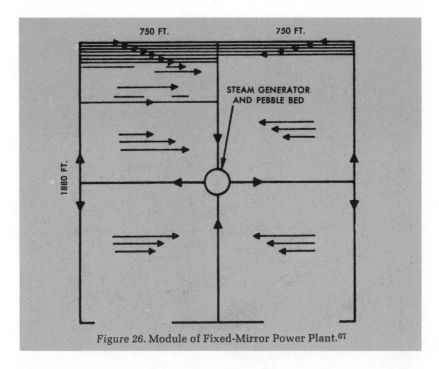

Figure 26. Module of Fixed-Mirror Power Plant.[67]

at 100 psi is heated in the collecting pipes by the focused sunlight and circulates through the pebble bed and/or the steam generator. Steam at 1000°F could be supplied from 9 of these modules to a centrally located turbogenerator of 162 megawatt electrical (MWe) capacity. Figure 27 illustrates the fixed mirror concentrator array. Costs of power from this facility is estimated to be competitive with alternative means of power generation.[67] Land costs would be negligible since, even at $1000/acre, the land cost is only $0.023/ft². Desert land is even cheaper.

A 1,000,000 MWe solar-thermal power plant has been proposed that would cover about 13,000 square miles of desert extending from the upper regions of the Gulf of California as far north as Nevada (Figure 28). The plant would use waste heat to produce 50 billion gallons of water each day, enough to meet the needs of 120,000,000 people. The proposed plant would use a circulating liquid metal (sodium or NaK) to extract heat from a solar farm and store it in a phase-change salt or eutectic mixture at temperatures

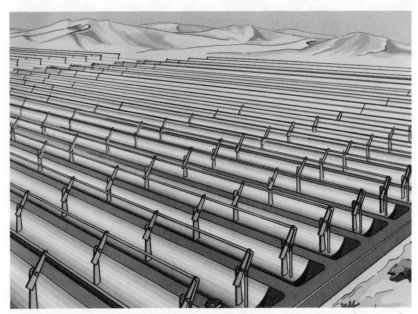

Figure 27. Artist's Concept of Fixed-Mirror Solar Concentrators Showing the Mirrors and the Tracking Heat Absorber Pipes.[67]

in excess of 1000°F. Power would be produced by a high pressure steam turbine-generator, and the low pressure steam from the turbine used to distill water. The total cost of solar heat collected by this plant is estimated at $0.50 per million BTU.[69]

Photovoltaic Power Generation

Solar cells offer a potentially attractive means for the direct conversion of sunlight into electricity with high reliability and low maintenance, as compared with solar-thermal systems. The present disadvantages are the high cost of about $25/watt,[70] and the difficulty of storing large amounts of electricity for later use as compared with the relative ease of storing heat for later use. The cost of solar cells is expected to be considerably reduced when cells are manufactured in large quantities using new production techniques for obtaining ribbons or sheets of single crystal silicon. At present

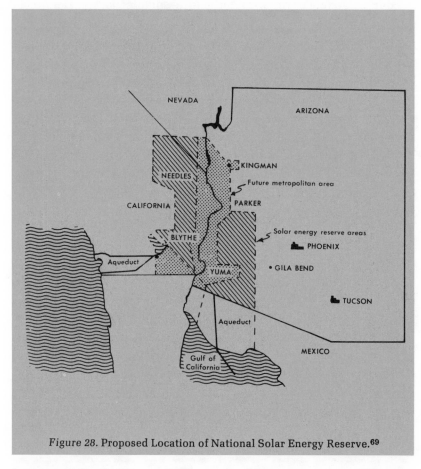

Figure 28. Proposed Location of National Solar Energy Reserve.[69]

large crystals of silicon or other semiconducting material are grown and then sliced into thin cells; new techniques for producing the thin slices directly use edge defined film growth,[71] dendritic growth,[72] rolled silicon,[73] or sheets of cast silicon that are recrystallized through heated or molten zones.[74] Silicon itself is very cheap since it is the second most abundant element in the earth's crust, and is produced in the U.S. at an annual rate of 66,000 tons at a cost of $600/ton, so when the most suitable of these mass manufacturing techniques is utilized the cost of solar cell arrays should be reduced to $1/watt or less, making them useful for the large scale generation of electric power.[71, 75]

Four companies which manufacture solar cells are Heliotech, Centralab, Solar Power Corporation (Exxon), and Sharp. Solar Power Corporation[76] sells a small solar power module that produces 1.5 watts at a solar intensity of 100 mW/cm². The current and power output characteristics of these solar cells (typical of solar cells in general) are given by Figure 29. Standard conditions are 0°C and

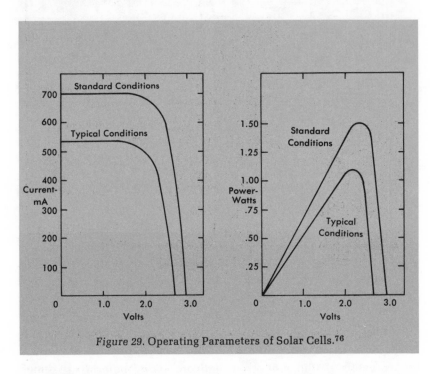

Figure 29. Operating Parameters of Solar Cells.[76]

1000 W/m² insolation, typical conditions are 25°C and 800 W/m² insolation. The solar array module consists of five 2.17 inch diameter silicon solar cells attached to a 13 ½ inch by 2.9 inch panel and is usually used to charge storage batteries to provide a continuous supply of power in remote locations. Tests in Arizona showed no degradation in output over a six month period. One power system being used at present to power navigational lights consists of 80 of these modules, 28 100 amp-hr 12 volt storage batteries, and the electronic control circuit. This power supply is cheaper to use than the alternatives; the Coast Guard saves about $3 million/year by using solar powered buoys.[77] The cost reduction is mainly due to the smaller number of trips out to the buoys for servicing. Wires are used to prevent seagulls from landing, but nothing is done about snow. NASA's experience testing solar cell arrays in Cleveland has shown no significant reduction in power due to dirt or dust accumulation and little problem with snow.

Large solar cell arrays have been considered for supplying the electric power needs of the western United States in 1990, assuming that solar cells can be mass produced at $1/watt. An array covering 192 square miles, coupled with pumped storage, would supply the 14,300 MWe needed by Arizona in 1990 for about $58 billion, and an array covering 2200 square miles (44 × 50 miles) would supply 40% of the electrical power needs of the 11 western states for a capital cost of around $637 billion.[78] Since these costs are far in excess of alternative means of power generation, it appears that even at $1/watt solar cells will be too expensive for central station power generation. The cost of solar cells must be reduced to about $0.20 per watt before solar cell arrays become practical for central station power generation.[79]

The cost of generating electric power with solar cells can be reduced by using concentrators to focus sunlight onto the cell. One simple type of concentrator is the reflecting cone (Figure 30). Without external cooling concentration ratios of up to five can be used without seriously reducing the cell performance due to cell heating. Higher concentration ratios are possible with external cooling. Solar cell arrays with concentrators must be steered to follow the sun; in the case of the conical concentrator the output is reduced below that with no concentrator if the angle of incidence is less than 60°, and at angles of less than 45° the output is negligible.[80]

Another related design is the channel concentrator consisting of two flat reflecting surfaces at an angle of 30° placed on both sides of a line of solar cells. The theoretical maximum concentration ratio is 3; an actual concentration ratio of 2.25 has been achieved with a channel concentrator array using 2 in × 2 in silicon solar cells at the base of the V channel.[70] Five channels with 30 cells each formed a 4.75 lb, 1 ft by 2 ft array producing 12 watts at 12 volts.

With external cooling, silicon solar cell outputs can be increased by more than 100 with concentrating systems.[81] Using experimental

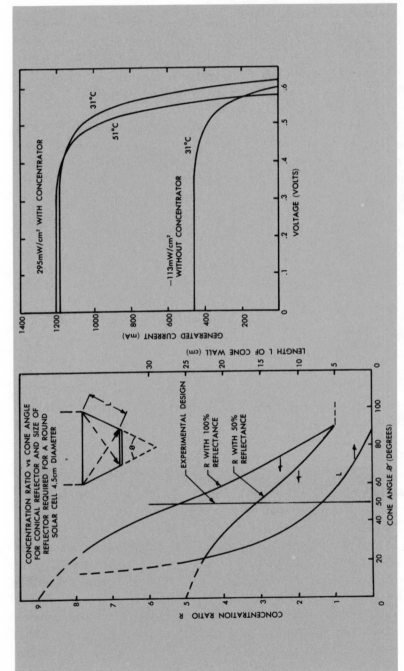

Figure 30. Concentration Ratio of Cone Reflector and Effect on Solar Cell Performance.[80]

data[82] for cells operating with solar fluxes between 14 and 28 watts/ cm^2, a system was designed to produce 50 watts of electrical power from 36 square centimeters of cell area by using a 5½ ft parabolic concentrator to provide a solar flux of 28 watts/cm^2.[83] The cells were water cooled to maintain their temperature at 200°F. Five watts were required to pump the water. A 250 watt electric power plant in the Soviet Union used a concentrator consisting of 26 plane mirror facets forming an approximate parabolic cylinder.[30] The concentrator increased the power output a factor of 5.2 over the power output with no concentrator, the solar cells were water cooled, and the overall plant efficiency was 2.7%.[84] Another plant was developed by the same group using channel concentrators with a concentration ratio of 2.5, and not requiring water cooling. These plants were developed to provide power for water pumps in the grazing areas of the southern regions of the U.S.S.R. One of them was installed at the Bakharden state livestock-breeding farm situated in the Kara-Kum Desert, Turkmenia. Its output equals about 400 watts, enough to lift from a depth of 20 meters a sufficient amount of water to water 2,000 sheep.[85]

Chapter 8

TOTAL ENERGY SYSTEMS

The feasibility of using solar energy to provide for all of the various energy needs of a home, business, or community requires either the development of inexpensive solar cells or an economical means of collecting solar heat at high temperatures and converting it to electric power. Photovoltaic cells can be combined with a flat plate collector (Figure 5) so that the radiant energy not converted into electric power is collected as heat and used to supply hot water, space heating, absorption refrigeration, and air conditioning. Figure 31 illustrates a solar cell flat plate collector which would permit utilization of up to 60% of the available solar energy. Collectors such as this mounted on vertical walls and/or part of the roof of a house or apartment building can supply all the various types of energy needs of the building. Figure 32 is a schematic showing the energy flows for a residential total energy system using solar cell flat plate collectors. This type of system is perhaps the ultimate in residential solar energy utilization, since both heat and electric power are produced without any moving parts, except for the pump or blower circulating coolant through the collector.

Advantages of this type of solar electric-thermal total energy system are: 1) The collector uses the same land area as occupied by the building, and thus there is minimal effect on the environment through use of land presently being used for other purposes, 2) About three times the present average household consumption of electric power can be collected from average-size family residences, even in the northeastern U.S. (This surplus energy could be used for charging an electric automobile), 3) The system is not vulnerable

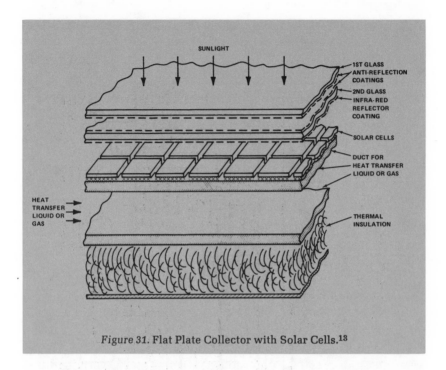

Figure 31. Flat Plate Collector with Solar Cells.[13]

to breakdowns in central energy generation stations or transmission systems, and 4) The small size of the individual unit makes proto-type testing and demonstration relatively inexpensive, and will help to attract consumer oriented industries.

Researchers at the University of Florida have built and tested solar water heaters, solar air heaters, a solar still, a five-ton solar air conditioner, a solar refrigerator, several solar ovens, a solar sew-age digester, solar cell arrays, several types of solar powered hot air engines, solar water pumps, a "solar-electric" car, and a solar house.[86] The solar house, occupied by a graduate student and his wife, uses solar energy for space heating, water heating, swimming pool heating, electricity, and recycling of liquid wastes with the solar still. A ⅓-horsepower hot air engine operating from a 5 ft parabolic concentrator drives a d-c generator to charge the solar-electric auto-mobile to provide pollutionless transportation from the solar house.[87] Thus it has been shown that it is technologically possible to use solar energy to provide all residential energy needs.

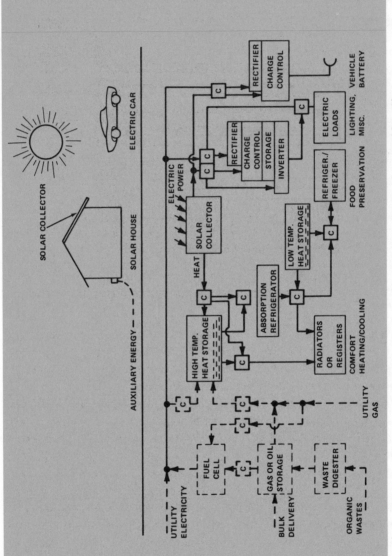

Figure 32. Schematic of a Solar Total Energy System for a Building.[13]

If inexpensive solar cells are manufactured, then the major remaining obstacle to the development of total energy systems is the problem of storing the electricity. One approach is to store the electricity in the form of hydrogen. Excess electric power generated during the day is used to electrolyze water to produce hydrogen and oxygen gas, which is compressed into storage tanks, and used in the evening with a hydrogen-oxygen fuel cell. This system is attractive in the long run, but too expensive at present for residential use.[88] Another possible energy storage medium is the flywheel. A new safe flywheel has been proposed with an energy storage capacity of 30 watt-hours per pound.[89] Excess electric power generated during the day is used to increase the rotational velocity of the flywheel, and in the evening the energy of the flywheel is used to generate electric power. Lead-acid batteries could be used, but if they were used to store a substantial fraction of all the electrical energy produced in the United States, it is questionable whether enough lead would be available.[90] Other electrochemical systems, however, might be possible, but more research needs to be done.

The other approach to developing a total energy system, not involving solar cells, is to collect the heat at a high temperature using a dynamic conversion system to produce electric power, and use the waste heat for space heating and cooling. One group studied four different types of total energy systems using concentrating collectors, high temperature heat storage, and a derated turbine, where the exhaust energy is used for heating and air conditioning. Another system with a flat plate collector driving an organic turbine generator was rejected as not being economically competitive with focused concentrator systems. They calculated the performance and economics of each proposed system for Albuquerque, N.M.[91-93] One day in four was assumed cloudy and the direct insolation taken to be 80% of the total. The cost of the residential solar energy systems were compared with a "normal" system supplying equal energy demands with utility electric power, and natural gas for space heating, air conditioning, and water heating. The results of these calculations indicate that solar total energy plants with high temperature collection and three levels of heat storage would be economically competitive with the "normal" system when the wholesale fuel cost reaches $0.90/MBTU.

Large users of energy such as apartment complexes, shopping centers, and industries can take advantage of solar-thermal total energy plants ranging in size from 0.2 to 20 megawatts. As of 1972 there were about 550 total energy plants in this size range in operation in the United States.[94] The more recently installed plants have averaged over five megawatts in capacity. Electrical storage problems for all types of total energy plants can be reduced considerably if the electric utility company owns and maintains these systems, and allows the excess power generated during the day to be fed back into the utility power grid. The electric power company could then give credit on the electric bill for power supplied by the customer. The major technical difficulty with this scheme is the phase-matching problem encountered when many different AC sources supply a common grid.

Chapter 9

INDUSTRIAL AND
AGRICULTURAL APPLICATIONS

The parabolic concentrator has provided an economical means of generating very high temperatures for small scale industrial applications and for research purposes, and solar heat at lower temperatures has been used for both industrial and agricultural drying operations. These represent two of the more promising commercial uses of solar energy.

Solar Furnaces

A megawatt furnace was built in France in the 1950's using heliostats to direct sunlight toward a large parabolic concentrator[64] (Figures 33 and 34). A similar 70-kilowatt furnace was built in Japan using a 10-meter-diameter parabolic concentrator. Another furnace of the heliostat type in Nantick, Mass. uses a spherical concentrator.

The Japanese furnace began operation in 1963 and produces temperatures in excess of 3400°C, the melting temperature of tungsten. Refractory bricks have been melted even in feeble sunlight. The furnace is used for studies of high temperature materials properties and some manufacturing. For example, alumina when melted in a graphite cylinder assumes a spherical shape because of its large surface tension. Turning the cylinder properly results in the formation of a fused alumina crucible which has much more desirable properties than a sintered one. Tungsten melted in an inert gas does not form a carbide even though the melting occurs on a graphite surface. Front surface aluminized mirrors used for the furnace showed a reduction in reflectivity from about 95% to 85% over a

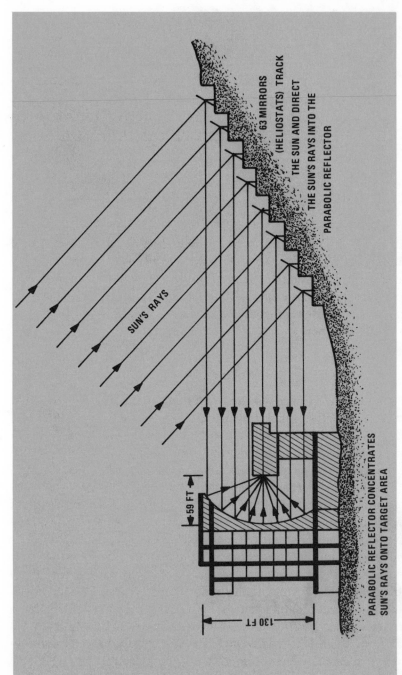

Figure 33. Schematic of 1000 kw Solar Furnace.

Figure 34. French Solar Furnace. (Courtesy of J. D. Walton)

five year period. At the present time all the mirrors are aluminized once each year.[95]

The French solar furnace is located in the Pyrenees at Odeillo-Font Romeu (altitude, 5900 feet), about 20 miles east of Andorra. At this location the sun shines as many as 180 days a year and solar intensities as high as 1000 watts per square meter are common. The solar furnace was completed in 1970 at a cost of about $2,000,000.

The parabolic reflector has a focal length of 59 feet, is 130 feet high and 175 feet wide and is composed of 9500 mirrors 17.7 inches by 17.7 inches. Since the parabolic reflector is too large to track the sun, 63 heliostats set in eight tiers are used to follow the sun and reflect its ray in parallel beams onto the parabola. The heliostats are 24.6 by 19.7 feet and each is composed of 180 mirrors 19.7 inches by 19.7 inches.

The solar energy incident on an area of about 23,000 square feet is concentrated by the parabolic reflector into an area about two feet in diameter. Sixty percent of the total thermal energy (about 600 kilowatts) is concentrated in an area one foot in diameter at the center of the focal plane of the parabola.

A similar but smaller solar furnace in the Soviet Union melts refractory materials at a temperature of up to 3,500°C, and is used for producing high purity refractories.[85]

Air Heaters

Solar air heaters have a great potential for improving agricultural drying operations around the world. At present a large portion of the world's supply of dried fruits and vegetables continues to be sun dried in the open under primitive conditions. Being unprotected from unexpected rains, windborne dirt and dust, and from infestation by insects, rodents and other aninmals, the quality is often seriously degraded, sometimes beyond edibility. In an increasingly hungering world, practical ways of cheaply and sanitarily preserving foods are needed.

There are two basic methods of solar dehydration. By the first method the necessary heat is supplied by directly exposing the material to solar radiation, which also enhances the proper color development of greenish fruits by allowing, during dehydration, the decomposition of residual chlorophyll in the tissue under direct solar radiation. The major drawbacks are the posible damage due to overheating, and relatively slow drying rates resulting from poor vapor removal in cabinet driers. According to the second method the foodstuff is heated by circulating preheated air. Since the drying material is not subjected to direct sunshine, caramelization and heat damage do not occur. A further advantage is that the circulating air absorbs the water vapor from the food, thus accelerating drying. On the other hand, products of inferior appearance may result if immature fruit is dehydrated, since shading prevents the breaking down of chlorophyll.

One dryer used a square meter area of steel chips beneath a glass cover to absorb solar radiation, and passed air to be heated through the chips. Steel chips are cheap, have a high heat transfer area per unit volume and excellent turbulence geometries, and an absorptivity of 0.97. Several agricultural products were dried and compared with an open air sun-dried control group. Peppers dried in the solar dryer had attactive bright colors as opposed to the brownish color of the slower drying control batch, which was sun-dried in the open. Similarly, for the dehydration of sultana seedless grapes, the sun dried control sample was rained upon, causing it to have a dark color. Soon afterward it was attacked by birds so the weighings were terminated. Raisins in the deydrator had a golden color and were dried in spite of continuous rainy weather.[96]

A variety of solar heaters have been developed for use in crop drying, space heating, and for regenerating dehumidifying agents. These various types of heaters provide air at 100°F above ambient with collection efficiencies of 50% or more. The heat transfer processes in air heaters are quite different from those in flat plate collectors which heat water. In the water-cooled collector, heat absorbed by the plate is transferred to the water tubes by conduction, so the absorber plate must have a high thermal conductivity. In an air heater the air can be in contact with the whole absorbing surface, so the thermal conductivity of the absorbing surface is of little importance. This makes solar collectors for heating air inherently cheaper than solar collectors for heating water. The main factors determining the efficiency of heat collection of a solar air heater operating at a given air inlet temperature are:

1. Heater configuration, that is, the aspect ratio of the duct and the length of duct through which the air passes.
2. Air mass flow rate through heater.
3. Spectral reflectance and spectral transmittance properties of the absorber cover.
4. Spectral reflectance properties of the absorber plate.
5. Stagnant air and natural-convection barriers between the absorber plate and ambient air.
6. Heat transfer coefficient between the absorber plate and the air stream.
7. Insulation at the absorber base.
8. Solar insolation.

V-corrugation of the absorber plate considerably improves the performance over that of collectors with flat absorbing surfaces. Spectrally selective coatings also improve performance. Air heaters of simple construction employing cheap materials have been shown to be capable of supplying air at temperatures above 150°F with good efficiency.[42] For crop drying only air temperatures below 180°F are needed.

One study of flat plate air heaters with two glass covers showed that if the air is passed between the two glass panes before passing through the blackened metal collector the outer glass temperature is reduced 4°F to 10°F, the collection efficiency increases 10% to 15%, and the temperature rise of the air is increased as much as 20%. Thus, it appears an attractive non-concentrating air heater design could use the two pass configuration and a V-corrugated absorber with spectrally selective coating.

Tests were made of air heaters with an absorber consisting of 96 parallel specularly reflecting aluminum fins 6.35 cm high, 0.635 cm apart, and 61 cm long. A single 0.317 cm glass coverplate was placed over the absorber, and air pumped between the fins. The collectors measured 61 cm × 61 cm. The collector with specularly reflecting fins was shown to be about 15% more efficient than an identical collector with diffuse fins.[97] Solar air heaters using hot water from water-cooled flat plate collectors have also been built.[98]

The use of concentrators to produce higher air temperatures for industrial operations, such as the 250° to 500°F needed by textile mills, has received little attention so far. Parabolic cylinder concentrators (Figure 6) or faceted concentrators such as proposed by Russell (Figure 27) could be used to heat air to high temperatures, and this heat can be inexpensively stored in rocks. The hot air can be used directly for textile or agricultural drying and other industrial operations requiring hot air at temperatures up to 1000°F.

Chapter 10

SOLAR STILLS

Solar stills are being used increasingly worldwide to produce drinking water from salty or polluted water. A still at the University of Florida is used to reclaim drinking water from household liquid wastes.[86] Solar stills are the cheapest means for desalting quantities of less than 50,000 gal of saline water per day in areas of reasonable sunshine, and production costs are currently about $3.50/1000 gallons.[99]

A solar still is typically a transparent plastic tent or glass enclosure containing a shallow pan of saline water with a black bottom. Sunlight heats the water in the pan, causing it to evaporate and recondense on the underside of the sloping plastic or glass and run down into collecting troughs along the inside lower edges of the transparent cover. The performance of solar stills has been calculated under various conditions of ambient temperature and insolation, and the results showed close agreement with data from a 4500 ft² solar still located at Muresk in Western Australia (Figure 35).[100] The daily output rose from about 0.1 lb/ft² (450 lb total) of water per day in the winter (July) to about 0.8 lb/ft² (3600 lb total) of water per day in the summer (December), so the range of production for this still is from 0.012 gallons to 0.1 gallons per day per square foot of collector. Similarly, a large 23,300 ft² solar still[101] on the island of Saint Vincent in the West Indies provides the most economical source of fresh water (other than rainwater), since underground natural sources are not available and the cost of shipping water to the island is high. The average daily output of the plant is about 0.05 gallons/ft² of collector, or more than 1000

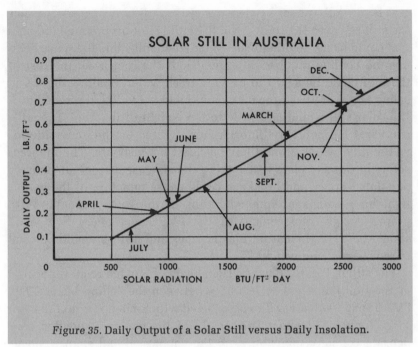

Figure 35. Daily Output of a Solar Still versus Daily Insolation.

gallons per day for the plant. Four mil polyvinyl fluoride film is used as the transparent cover.

For buildings solar stills may be mounted on rooftops (Figure 36).[102] An advantage of this approach is that the cost of the solar

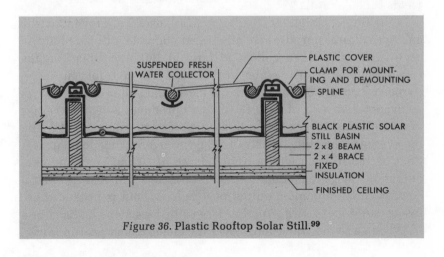

Figure 36. Plastic Rooftop Solar Still.[99]

still is partially offset by the savings in ordinary roof costs, since the still replaces the roof. Also, the still is not occupying land area that could be put to other uses. Since solar collectors for space heating and cooling require only about half the roof area, the rest of the roof could be a still to provide fresh water. Shallow depths of saline water are used for maximum yield, so unlike the roof pond, the solar still would add little to the weight of the roof. The stills can use 4 mil polyvinyl fluoride film, treated on the underside for wettability, which should have at least a 5-year life. The inherent safety hazards of glass restrict its extensive use on roofs in densely populated areas, and breakage could cause puncture of the watertight liner with subsequent flooding of the room beneath. With proper design, replacement of the PVF cover every five years should be a simple matter. Plastic pipes and fittings would be used to reduce cost and weight.[102]

The still in Figure 36 uses a rigid basin of molded resin supported directly on the 2 × 4 inch braces between the ceiling beams. The PVF transparent cover film is fastened with S-clamps onto the main 2 × 8 inch roof beams. The weight of the center-suspended condensate collector contributes to the vapor seal and shapes the V-cover so that distillate drains to the collector. The result is an inexpensive waterproof roof which provides a supply of fresh water. Accidental cover damage would, at worst, allow rain to drain into the condensate collector. PVF covers have produced the highest yields for solar stills.[103]

In the Soviet Union large solar stills are used both for industrial and agricultural purposes. A reinforced concrete still was built in 1970 in the Shafrikan collective farm at Bukhere Oblast in the Uzbek Republic.[104] The water in that area was unusable for many purposes because of its very high mineral and sulfur content. With an evaporative area of 6500 ft², the still produces a yearly average of 0.08 gallons/ft² per day, a total of about 540 gallons per day on the average. The still consists of 39 glass covered independent sections of 168 ft² each with a trough depth of 10 cm. The maximum output of 80 gallons/hour occurs between 2 PM and 4 PM (in August) and a minimum output of 5 gallons/hour is produced between 3 AM and 7 AM. Another large still uses steps inclined at a 2°–3° angle so the water flows over the steps, from upper to lower, until it reaches the discharge drain. This flow enhances evaporation and increases the output and efficiency of the still about 20%.[105]

The Krzhizhanovsky Power Institute in Moscow has also been studying various aspects of solar stills. Theoretical studies were conducted of heat and mass transfer processes in solar stills of the hotbox type and techniques for calculating the performance of these stills were developed. In a properly designed still most of the solar energy that passes through the glass (or film) is used to evaporate saline water. As a result, the space within the still is filled with a steam-air mixture. The energy balance conditions during operation of the still are such that the surface of the glass is at a lower temperature than that of the steam-air mixture, with the result that water vapor condenses on the glass surface, whereas the condensate runs down the inclined glass, drips into the groove and is collected in the tank. Well instrumented solar stills were constructed to investigate these processes. During the tests the temperature of the water heated by the sun varied from 74°F–207°F while the temperature of the glass condensing surface varied from 61°F–192°F. As a result of these studies equations were developed which accurately describe heat and mass transfer processes in this type of solar still.[106]

The effect of wind speed and direction on the output of a solar still of the greenhouse (glass) type was studied by using a fan to blow air across a small still. For saline water temperatures of 104°F, 131°F and 158°F the wind speed was varied from 0–26 ft/sec at wind directions of 0, 45, 90, 135 and 180°; and for all wind directions and temperatures the maximum still output was achieved for a wind velocity of about 16 ft/sec. Increasing the wind velocity up to this value increases the rate of heat removal from the glass cover, which increases the rate of condensation on the glass, resulting in an acceleration of the evaporation process and as much as a 25% increase in still output. Further increases in wind speed lead to a reduction in the saline water temperature which reduces the evaporation rate and still output. The most favorable wind direction is parallel to the condensing surfaces (90° angle).[107]

Chapter 11

CLEAN RENEWABLE FUELS

Most of the energy used in the United States today comes from fossil fuels produced many years ago from solar energy. Clean renewable fuels to supplement and eventually replace these fossil fuels can be produced from plant life grown under more optimum conditions than found in nature, and from organic waste materials. The various processes for the production of these fuels listed in Figure 37 are aimed at converting organic materials with a low heating value per unit weight into higher heating value fuels similar to the fossil fuels currently in use. Another possible technique is the use of high temperature heat from solar concentrators to operate a regenerative thermochemical cycle for the production of hydrogen; the hydrogen can be used directly or utilized for the production of hydrocarbon fuels such as methane.

Perhaps the oldest and simplest technique for the production of a clean renewable fuel is to grow plants and burn the plants for energy; this could be done on a large scale for electric power generation.[108] Air pollution from such a plant is minimal since virtually no oxides of sulfur are produced, particulate emissions can be controlled with precipitators, and the CO_2 released is reabsorbed by the growth of new plants. Up to 3% of the incident solar energy can be absorbed by plants, and this energy is released when the plants are burned.[109] For a 1000 MWe steam-electric power plant operated at a load factor of 75% with a thermal efficiency of 35%, 150 square miles of land area is required to fuel the plant if the average insolation is 1400 $BTU/ft^2/day$ and the capture efficiency of the plants is 3%. It has been calculated that the total cost of the

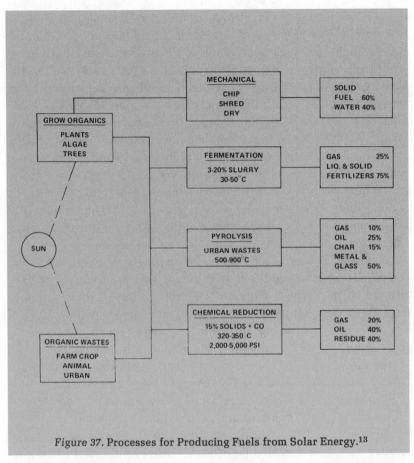

Figure 37. Processes for Producing Fuels from Solar Energy.[13]

fuel would be $0.06/MBTU for a $250/acre land cost, 1400 BTU/ft²/ day insolation, 3% capture efficiency, 8% interest rate, 0.6% tax rate, and $200/acre harvesting cost, and the total cost of the electric power from this "energy plantation" is computed to be 5 mills/ KWh, based on a $200/KW capital cost and 28 year life for the power plant.[108] A "worst case" fuel cost was determined to be $0.40/MBTU if the capture efficiency is reduced to 1% and the harvesting cost increased to $700/acre, which results in a power cost of 8.5 mills/KWh. The annual operating cost is taken to be $2 million/year, and insurance and tax costs 0.12% and 2.35% of the capital cost of the power plant. This study concluded that this type of plant would cost no more to build and maintain than a conventional fossil fuel steam electric plant and that the energy plantation is a renewable resource and is an economical means of harnessing solar energy. It is not at all obvious at the present time what type of plant (trees, grasses, etc.) will result in the lowest power costs. The NSF/NASA Solar Energy Panel concluded that using trees the fuel cost at the power plant might range from $1.50 to $2/MBTU.[13]

Some power can also be produced by the combustion of organic wastes, which also reduces problems of disposal of these wastes. It has been estimated that the total animal and solid urban wastes which can be collected at reasonable cost could provide about 6% of the heat energy requirements for electric generating plants. The most promising use of solid animal wastes is in connection with large feedlot operations where large quantities are accumulated at one location and disposal presents a continuing problem.

Anaerobic fermentation of organic materials results in the production of methane and carbon dioxide. This process can be used (Figure 38) to convert from 60% to 80% of the heating value of organic materials into methane, which can serve a wide variety of uses including powering automobiles. Methane can also be used in existing natural gas pipelines. Algae grown in sewage ponds can be used for the production of methane; costs of producing methane by this method are estimated between $1.50 and $2/MBTU.[13]

Pyrolysis has also been used for many years to convert organic materials to gaseous, liquid and solid fuels. Any organic materials can be used, and in addition plastics, rubber products, and other similar materials can also be used. The gases produced are a mixture of hydrogen, methane, carbon monoxide, carbon dioxide, and hydrocarbons. About two barrels of oil can be produced per ton of dry organic material. A plant handling 1000 tons of waste per day (Figure 39) could dispose of the solid wastes produced by a city of 600,000 people.

At temperatures around 600°F and pressures between 2000 psi and 4000 psi organic materials can be partially converted into oil. In laboratory tests oil yields up to 40% of the weight containing about ⅔ of the heating value of the initial dry organic matter have been obtained.

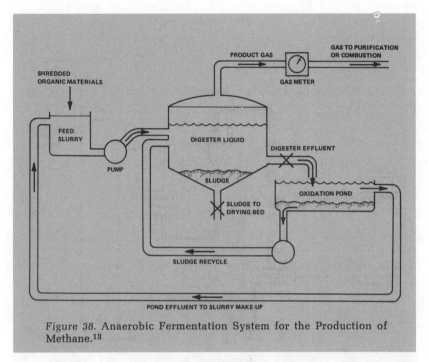

Figure 38. Anaerobic Fermentation System for the Production of Methane.[13]

Hydrogen may be thermochemically produced directly from water using solar heat. For example, one regenerative chemical cycle operates with bromides of calcium and mercury in a four step process with a maximum temperature of 1350°F.[110] The four reactions are

1) $CaBr + 2H_2O$ $\rightarrow$ $Ca(OH)_2 + 2HBr$ 1350°F
2) $Hg + 2HBr$ $\rightarrow$ $Hg Br_2 + H_2$ 480°F
3) $HgBr_2 + Ca(OH)_2$ $\rightarrow$ $CaBr_2 + HgO + H_2O$ 400°F
4) HgO $\rightarrow$ $Hg + (0.5)O_2$ 900°F

The net result of these four reactions is:

$$H_2O \rightarrow H_2 + (0.5)O_2$$

Water is thus separated into hydrogen and oxygen at temperatures easily obtainable by linear concentrators. The hydrogen and oxygen are released at separate points in the cycle, and the chemicals used are regenerated permitting virtual 100% recovery of the chemicals

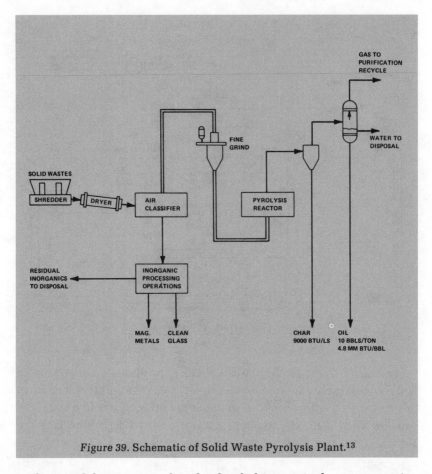

Figure 39. Schematic of Solid Waste Pyrolysis Plant.[13]

without sideloops. One drawback of this particular regenerative process is the large amount of materials circulated per unit product. This cycle is an example of a large number of regenerative thermochemical cycles that have been proposed for the production of hydrogen with temperatures obtainable on a large scale with solar concentrators.[111]

Figure 40 illustrates the relative 1972 cost of solar-produced clean renewable fuels and fossil fuels. The costs of fossil fuels have risen considerably since 1970, so solar synthetic fuels are going to continue to become increasingly competitive.

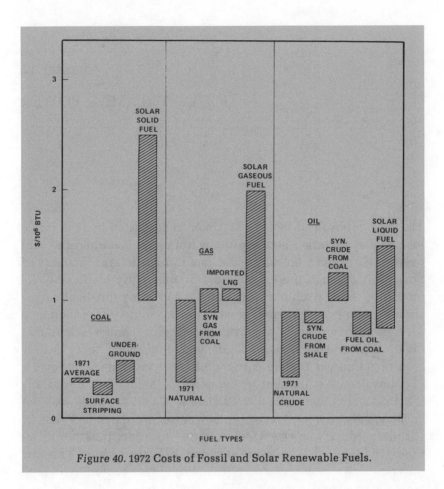

Figure 40. 1972 Costs of Fossil and Solar Renewable Fuels.

Chapter 12

OCEAN THERMAL POWER

The French physicist Jacques D'Arsonval suggested in 1881 that a heat engine operating between the warm upper layer and the cold deep water of the tropical oceans could produce large amounts of power.[112] Although the engine must be inherently inefficient, the amount of heat available is enormous, and since this heat comes from the sun ocean thermal power is appropriately classified as a form of solar power. D'Arsonval suggested a number of possible high vapor pressure working fluids, including ammonia.

In 1929 Georges Claude, a friend of D'Arsonval, demonstrated a 22-kilowatt ocean thermal power plant in Mantanzas Bay, Cuba (Figure 41), but due to its low efficiency ($<1\%$) the plant was not economically competitive with other power plants at that time.[113] Claude used surface sea water admitted to a low pressure evaporator to provide low pressure steam to drive the turbine. This low pressure steam was then recondensed by direct contact with cold seawater in a spray condenser. The Claude cycle avoided large heat exchangers required by closed cycle plants to vaporize and recondense a high vapor presure working fluid, but did require a large turbine of inherently low efficiency. The relatively high vacuum required maintaining large leak-tight connections and the removal of dissolved gases from the water. The plant itself was located on land and 2-km long tubes brought cold water from the depths, with resulting heating of the water as it flowed through the tubes. In spite of the economic failure of the project, Claude's plant was the first to demonstrate power generation from ocean temperature gradients.

Two large experimental power plants of 3.5 MWe each using the

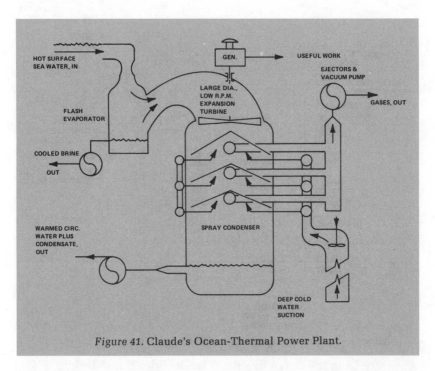

Figure 41. Claude's Ocean-Thermal Power Plant.

Claude cycle were built by the French at Abidjan off the Ivory Coast in 1956 to utilize a thermal difference of 36°F. An 8 ft diameter pipeline was built extending to a depth of 3 miles about 3 miles from shore, but difficulties in maintaining this pipeline prevented the plant from operating at full capacity. About 25% of the power generated was required for the pumps and other plant accessories. The plants were finally abandoned.

Two approaches to improving the Claude cycle are the use of controlled flash evaporation[114] and the indirect vapor cycle.[115] The controlled flash evaporation system (Figure 42) eliminates major problems of deaeration and seawater corrosion associated with the Claude cycle and produces fresh water in addition to electric power. The flash evaporater consists of a large number of parallel vertical chutes with films of warm seawater flowing down. As the pressure drops, water evaporates and the vapor flows downward. This low pressure steam then flows through the large low pressure turbine and into the condenser where it is cooled and condensed by cold

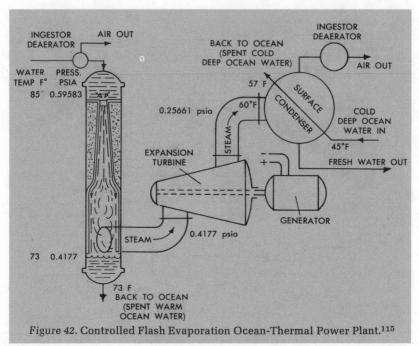

Figure 42. Controlled Flash Evaporation Ocean-Thermal Power Plant.[115]

seawater from the ocean depths. If fresh water is not desired, steam from the turbine can be condensed by direct contact with cold seawater (as in the Claude cycle) with a slight increase in power output. Deaeration in this cycle is accomplished at a low cost with practically no power requirement. About 11.5 gallons of pure water can be produced per 1000 gallons of warm water circulated. This system still suffers from the large, inherently inefficient low pressure steam turbine.

The indirect vapor cycle requires the addition of a boiler, but permits the use of higher pressure working fluids with a much smaller and more efficient turbine (Figure 43). Since the efficiency of ocean thermal plants can be only about one-tenth that of modern steam plants, the amount of heat transferred in the boiler and condenser per unit power output must be about ten times as large. It does not follow, however, that the costs of these components will be ten times as great. Since ocean thermal plants will operate at relatively low pressures and ambient temperatures, the tube walls can be thinner and cheaper materials can be used, so the cost per unit of

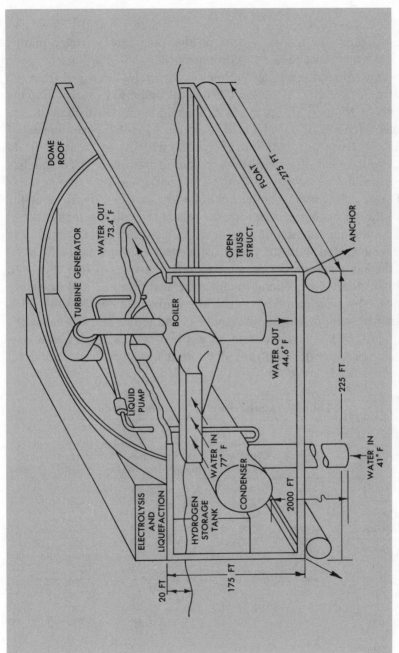

Figure 43. Floating Ocean-Thermal Power Plant.[115]

heat transfer should be much less for ocean thermal boilers and condensers than for those used in high temperature steam plants.

A floating power plant has been proposed that could use propane as the working fluid in an indirect vapor cycle. Seawater from the warm surface layer is passed through the boiler to vaporize propane at about 150 psi. The propane exhausted from the turbine is condensed at about 110 psi by cold seawater. In 1965 the capital cost of this plant was estimated at $168/KW, which was comparable to the capital cost of a fossil-fueled plant at that time.[116] To equalize pressure differences in the boiler and condenser, the plate heat exchanger acting as the boiler could be lowered to a depth of 290 ft and the plate condensers lowered to 150 ft, with the turbines and other components at intermediate depths. A modular design has been suggested with the boiler, condenser and engine modules all of the same standard size, such as 8 ft $\times$ 8 ft $\times$ 40 ft (Figure 44), which should reduce manufacturing, transportation, and assembly costs. The plant would be neutrally buoyant at the depth which minimizes the pressure differences in the boiler and condenser.[117]

Figure 45 is a schematic of an indirect cycle system and a generalized temperature entropy diagram; characteristics of potential working fluids are given in Table 8.[118] The ideal cycle efficiency is

Table 8. Comparison of Working Fluids[118]

Fluid	Ideal Cycle Efficiency (%)	Cycle Efficiency (5% $\Delta P/P$)	High Pressure (psia)	Low Pressure (psia)	Pump Work (kw)	Ideal Mass Flow (lb/min)
Ammonia	3.72	2.71	118	81	1079	317,600
Butane	3.82	2.81	29	20	859	976,000
Carbon Dioxide	2.89	1.67	799	609	36,033	2,873,000
Ethane	3.90	2.04	53	411	25,300	1,495,000
R-12	3.68	2.57	78	56	2,450	2,630,000
R-22	3.68	2.54	126	91	3,200	1,978,000
R-113	3.65	2.91	5	3	170	2,436,000
R-500	3.67	2.55	92	66	2,750	2,205,000
R-502	3.61	2.41	140	103	4,552	2,756,000
Propane	3.67	2.46	115	85	3,706	1,084,000
Sulphur Dioxide	3.72	2.82	45	30	634	1,041,000
Water	3.78	3.26	0.3	0.15	1.4	155,500

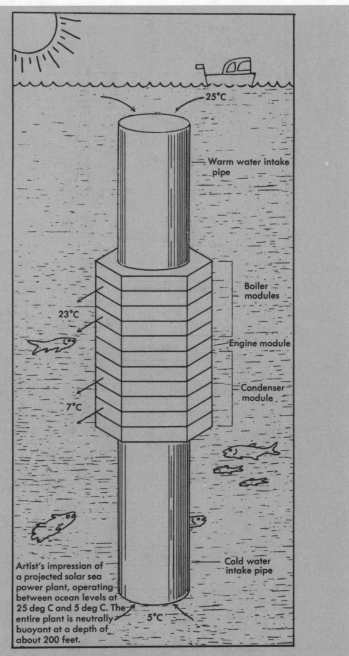

Warm water intake pipe

Boiler modules

Engine module

Condenser module

25°C

23°C

7°C

Cold water intake pipe

Artist's impression of a projected solar sea power plant, operating between ocean levels at 25 deg C and 5 deg C. The entire plant is neutrally buoyant at a depth of about 200 feet.

5°C

Figure 44. Modular Ocean-Thermal Power Plant.[115]

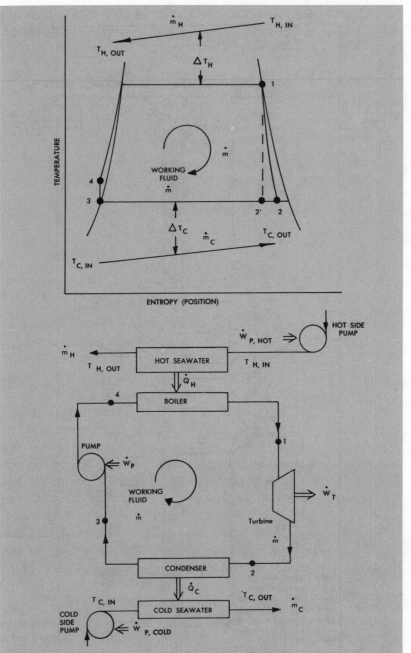

Figure 45. Schematic and T-S Diagram for Ocean-Thermal Power Plant.[115]

based on a maximum cycle temperature of 65°F and a minimum cycle temperature of 45°F. Ammonia is the best working fluid from the heat transfer standpoint.

Table 9. Heat Transfer Coefficients of Working Fluids[118]

Fluid	Relative (Condensing)	Relative (Boiling)
Ammonia	1	1
Butane	0.15	0.32
Carbon Dioxide	0.18	0.18
Ethane	0.11	0.21
R-12	0.11	0.11
R-22	0.16	0.14
R-113	0.09	0.13
R-500	0.12	0.13
R-502	0.10	0.11
Propane	0.13	0.27
Sulphur Dioxide	0.38	0.33
Water	0.92	2.27

Various heat exchanger geometries are illustrated by Figure 46. For ammonia a single stage turbine with a 7 ft wheel diameter could generate 25 MW at 1800 rpm; for propane a 12 ft wheel diameter single stage turbine could produce 30 MW at 600 rpm. Propane and ammonia appear at present to be the most attractive working fluids.

The amount of energy available for ocean thermal power generation is enormous, and is replenished each year as the sun heats the surface layers of oceans and melts snow in the arctic regions causing cold currents to flow deep beneath the surface toward the equator. It has been estimated that the tropical oceans in the year 2000 could supply the whole world with energy at a per capita rate of consumption equal to the U.S. per capita rate in 1970 and suffer only a one-degree C drop in temperature. Also, if nutrient-rich cold water is brought from the ocean depths and released near the surface, this could result in a substantial increase in fish populations, as occurs naturally off the coast of Peru. Another advantage could be a slight lowering of tropical temperatures. At a depth of 1300 ft 30 miles from Miami the temperature is 43°F, as compared with a surface temperature from 75°F–84°F, so this could be a good location for an ocean-thermal power plant.

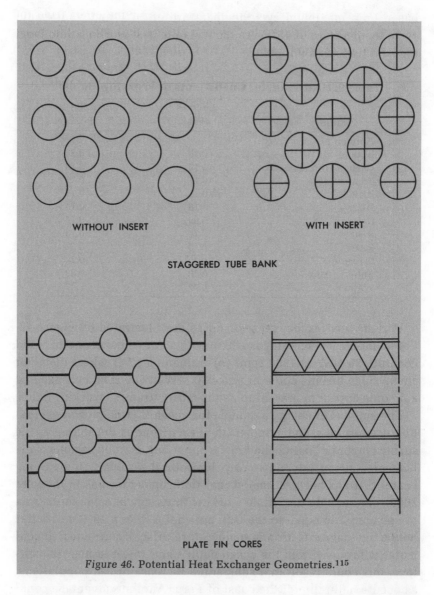

WITHOUT INSERT   WITH INSERT

STAGGERED TUBE BANK

PLATE FIN CORES

Figure 46. Potential Heat Exchanger Geometries.[115]

A "sea plant" has been proposed with a floating propane cycle ocean-thermal electric power plant, a separate ocean-thermal flash evaporation plant for producing pure water, and various chemical industries based on extracting oxygen and raw materials from the ocean.[119] Noting that the Gulf Stream alone could supply 200 times the total power requirements of the United States, the cost of a 100 MWe plant is estimated at $20 million ($200/KWe) and the cost of fresh water at $0.04/1000 gallons. This cheap power and cheap water makes possible a variety of energy intensive chemical process plants. Oxygen gas, extracted from seawater, could be liquified using propane turbines to drive the refrigeration compressors, and cold water from the ocean can be used as a convenient heat sink at lower than usual ambient temperatures. Chemical plants using raw materials extracted from seawater would benefit from the cheap power. Bromine and magnesium are already being produced commercially from seawater.[120] In addition, one of the best ways to transmit power to shore may be to electrolyze water to produce hydrogen and oxygen, and then liquify these gases, which can then be shipped or piped to shore.

The capital cost of sea thermal power plants is compared with that of other types of power plants (Figure 47). Although fossil power plants tend to cost less, the cost of delivered power can be higher because of fuel costs, which are negligible for solar, sea-thermal, geothermal, and wind plants. These cost ranges are subject to considerable change as technology advances and economic conditions change.

Another possible sea-thermal conversion system, which does not use any working fluid, is the Nitinol engine developed by researchers at the Lawrence Berkeley Laboratories. The Nitinol wires bend and straighten as they are immersed in cool and warm water, respectively. The force they exert during this process is converted into rotational motion, and small amounts of electric power have been generated by a laboratory model. Since this engine operates between temperature differences of only a few degrees it may eventually be useful for sea-thermal power systems. The small laboratory model is reported to operate at 70 rpm with a temperature difference of 41°F to produce 0.23 watts with an absolute thermal efficiency of 3.4%.[121]

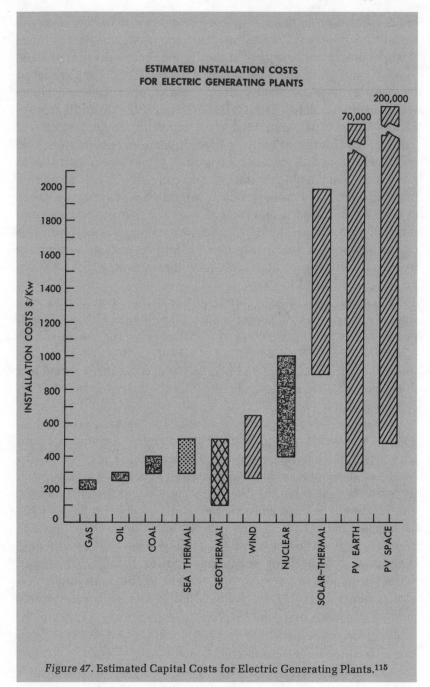

Figure 47. Estimated Capital Costs for Electric Generating Plants.[115]

Chapter 13

GEOSYNCHRONOUS POWER PLANTS

The concept of placing a large solar array in geosynchronous orbit and transmitting this power to earth was proposed in 1968,[121, 122] and since has received increasing attention as a potential major energy resource for the next century. The basic motivation for placing a solar array in space is the increased availability of solar energy in space, as illustrated by Table 10. Up to 15 times as much solar

Table 10. Average Availabilities of Solar Energy[123]

Availability Factor	Average On Earth	In Synchronous Orbit	Average Ratio
Solar Radiation			
Energy Density	0.11 watts/cm²	0.14 watts/cm²	4/5
Percentage of Clear Skies	50%	100%	1/2
Cosine of Angle of Incidence	0.5	1.0	1/2
Useful Duration of Solar			
Irradiation	8 hr	24 hr	1/3
Product			1/15

energy is received by a solar array in space as the same array would receive on the ground, and this energy is received continuously, nearly 24 hours a day. Now that NASA is developing the space shuttle to permit the routine exploitation of the space surrounding the earth, the economics of geosynchronous power plants are becoming more attractive.

The basic concept is illustrated by Figure 48. Concentrators would reflect sunlight onto an advanced, lightweight solar array.

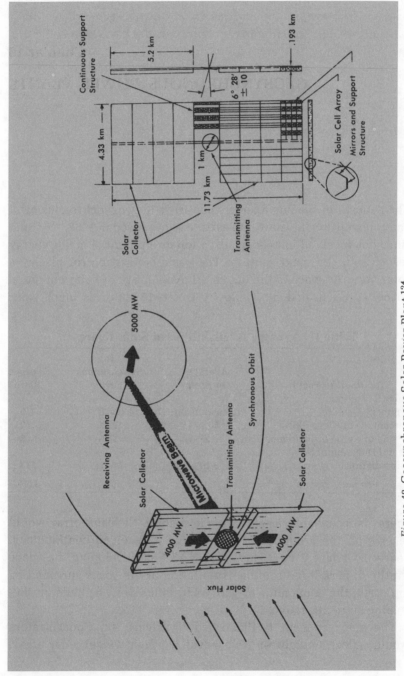

Figure 48. Geosynchronous Solar Power Plant.[124]

The two symmetrically arranged collectors convert solar energy directly to electricity which powers microwave generators within the transmitting antenna located between the two large collecting panels. The 1 km diameter antenna transmits the power to a 7.4 km diameter receiving antenna on the ground (Figure 49) with an over-

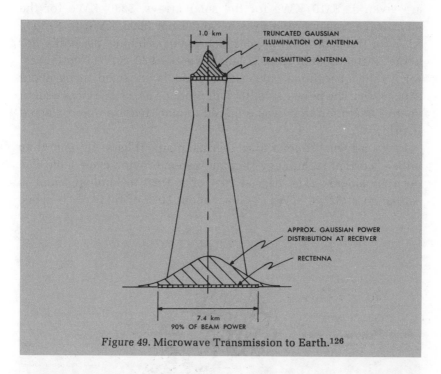

Figure 49. Microwave Transmission to Earth.[126]

all efficiency of about 68%. The microwave transmission system is expected to cost about $130/KWe.[125] To achieve the necessary coherent transmission, the many separate elements of the transmitting antenna are phase locked onto a pilot signal originating from the center of the receiving grid, so that it is impossible to direct the beam away from the receiving antenna. Since the receiving grid does not block sunshine, the land beneath can be used for growing farm crops. Microwave intensities reaching the earth are safe.

The solar cells in the array are projected to have an 18% efficiency, 2 mil thickness, and cost $0.28 per cm², which should lead to a 430 watt per pound array costing $0.68 per cm² and having a 30 year life. The array is expected to suffer a 1% loss of solar cells from micrometeoroid impacts over a 30 year period. Cost estimates for a small several hundred megawatt prototype plant, based on current shuttle cost estimates and near-term solar cell technology, are given as $310/KWe for the solar arrays, $230/KWe for the microwave transmission system, and from $800/KWe to $1380/KWe for transportation to geosynchronous orbit and assembly, for a total system cost of from $1340/KWe to $1920/KWe. Capital cost for a fully operational 5000 MWe plant is expected to be about $800/KWe. The power satellite will produce more energy in its first year of operation than was required to manufacture it and place it in orbit.[125]

Large geosynchronous solar-thermal plants (Figure 50) operating with a "current technology" helium/xenon brayton cycle have also been considered. The capital cost of a 1980 technology plant is estimated at $2540/KWe.[127] Since about 80% of this cost is space

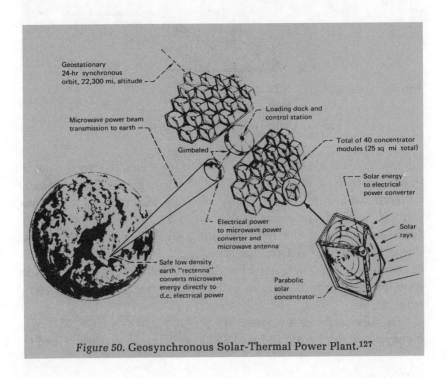

Figure 50. Geosynchronous Solar-Thermal Power Plant.[127]

transportation, this cost should be reduced if a fully reusable space shuttle becomes operational and lighter weight reflecting surfaces become available. Based on the same projected 1980 technology advanced solar cell systems are estimated to cost $2950/KWe, slightly more than the solar-thermal system. Another estimate for the capital cost of solar cell geosynchronous plants that was given lies in the range of $1400/KWe to $2600/KWe.[128] An earlier estimate of $2100/KWe was given for a prototype solar cell synchronous power plant based on a study by the A. D. Little/Grumman/Raytheon/Textronics team.[129] This group has been conducting a study of the solar cell synchronous power satellite for several years.

Chapter 14

POWER FROM THE WIND

A small fraction of the solar energy falling on the earth each day is converted into surface winds that are quite strong and in some areas provide a useful source of energy for performing mechanical work and generating electric power. Although windmills have been used more than a dozen centuries for grinding grain and pumping water, interest in large scale electric power generation has developed over the past 50 years. In 1920 J.B.S. Haldane wrote: "Ultimately we shall have to tap these intermittent but inexhaustible sources of power, the wind and sunlight. The problem is simply one of storing their energy in a form as convenient as coal or petrol. . . . During windy weather the surplus power will be used for the electrolytic decomposition of water into oxygen and hydrogen. These gases will be liquified and stored in vats, vacuum jacket reservoirs, probably sunk in the ground."[131]

In 1939 work was begun on a 1.25 MWe wind power plant on Grandpa's Knob near Rutland, Vermont. Electricity was generated and delivered to the utility transmission grid in October, 1941, the first synchronous generation of power from the wind. The rotors and electric generator were mounted on a 110 ft tower and turned in any direction to face the wind. The two stainless steel blades weighed 7.5 tons each and swept a circle 175 ft in diameter with a rated speed of 28.7 rpm. Full power operation was achieved for wind velocities in excess of 30 miles per hour, which occurred about 70% of the time. Icing of the blades was not a problem since the ice would break up during rotation. The total weight of the wind power generator was 250 tons, and the cost was slightly over one million dollars.[132]

Partly because the project was rushed to completion in the days preceding World War II, it was plagued with component failures. Replacements were especially difficult to obtain during the war. Finally, on March 26, 1945, the blade broke during a storm. Because of the limited financial resources of the company operating the plant the generator was not repaired, but dismantled and removed from the site.[133]

Based on the experience at Grandpa's Knob, the Federal Power Commission conducted a study of wind electric power generation for use with interconnected utility networks, and concluded that a power plant capacity between 5 and 10 Megawatts could make wind power economical. A 7.5 MWe unit was designed using two-bladed propellers, similar to the propellers used on small airplanes. A separate design for a 6.5 MWe plant used three-bladed propellers. A wind-driven d.c. generator provided power to a converter which produced synchronous a.c. power. The projected costs of these plants, in 1945 dollars, were $68 per kilowatt of capacity for the 7.5 MWe unit and $75 per kilowatt for the 6.5 MWe unit.[134]

In addition to the Grandpa's Knob experiment, a variety of similar projects have been undertaken around the world. A 100 KWe direct current wind turbine with a 30 meter propeller diameter was operated in the Soviet Union in 1931. In 1942 a three bladed propeller 50 KWe a.c. plant was operated in Germany, and the next year a 20 KWe generator using two 6-bladed 9-meter diameter propellers was tested in Berlin. In Denmark, a 200 KWe generator with a single 3-bladed propeller of 24-meter diameter has been operated, and a 100 KWe 15 meter 3-bladed turbine has been tested in England. Between 1961 and 1966 a 35 meter diameter, 100 KW double-bladed wind turbine operated in West Germany, following tests with a 10 KWe model. The power output of the 100 KWe unit increased linearly from 10 KWe at a wind velocity of 4 meters/sec to 90 KWe at 9 meters/sec. The power output was usually held to 90 KWe for wind velocities greater than 9 meters/sec. The most spectacular European wind generator has been the 31 meter diameter, 800 KWe generator built in France in 1958. This generator used a single 3-bladed propeller.[135]

At present a company in Switzerland is manufacturing 5 kilowatt wind generators with a 5 meter diameter propeller at a cost of $1900, plus $200 freight for delivery to the United States, with delivery in about 6 weeks. A 400 watt generator is also on sale in Germany, and a 1 KWe generator is made in Italy.[136]

One system for a house in Maine uses a 2 KWe wind generator manufactured in Australia with 19 storage batteries and a small dc-ac inverter to provide all the electric power needs for the home, including power for lights, household appliances, power tools, and the television.[137] The storage batteries provide enough reserve power for four days without wind; a gasoline generator is used as an emergency backup system in case of prolonged calm periods. The only maintenance required for this system is a change of one quart of oil in the gearbox once every five years.

The wind generator uses a 12 ft diameter propeller and produces up to 2 KWe of 115 volt dc power. The voltage regulator panel is incorporated as part of the wind generator. The transistorized voltage control lowers the charging rate when the output voltage exceeds the voltage of the batteries in the fully charged state. An anemometer on the wind generator measures the wind speed in miles per hour. The nineteen 130 ampere-hour batteries are connected in series to provide 15 KWH of storage at 115 volts. The lights and many of the appliances including the vacuum cleaner, electric drill, skill saw, sewing machine, and water pump are operated on dc. Only the television and stereo require ac, which is provided by a small inverter.

This system costs about $2800, including $1600 for the wind generator, $700 for the batteries, $100 for the small inverter, and $200 for wiring and other small components. The cost of the electric power for this home is about 15 cents per KWH—about half the cost of power from a gasoline or diesel plant, but much higher than the power company rate. This house was located 5 miles from the nearest paved road and the local power company would have charged $3000 to run a line to the house, so the wind power system was the most economical source of electric power for this home.

For those requiring more power, a company in Switzerland sells a 6 KWe wind generator complete with automatic controls, for $3000.[137] This unit has a three-bladed propeller of 16½ ft diameter and can operate completely automatically and unattended in winds as high as 150 mph. Full output is delivered in a 25 mph wind. Solid state inverters are also available, for example a 2 KWe dc-ac inverter sells for about $1600.

For commercial electric power generation, wind generators of power outputs in excess of 100 KWe will be needed. No storage is required for the utility network as long as the average hydroelectric generation is several times greater than the power provided by wind generators, because the output of the hydroelectric plants can be varied to compensate for changes in wind power output. Also, the use of large numbers of wind generators at separate sites will tend to reduce storage requirements and smooth out short-term fluctuations in total wind power output.

The largest wind generator built in recent times was the 800 KWe unit operated in France from 1958–1960. The flexible three-bladed propeller was 105 ft in diameter and produced the rated power in a 49 mph wind with a rotational speed of 47 rpm. The maximum power delivered was 1.2 MWe. After 18 months in operation the flexible propeller was replaced by a rigid one in an effort to improve the performance of the generator. One of the blades of the new propeller soon broke and the hub of the machine was torn up by the unbalanced forces acting on the propeller. The generator was never operated again.[138]

A 100 KWe generator will soon be built by NASA with NSF support at the Lewis Research Center Plum Brook test facility at Sandusky, Ohio. This will be the largest wind generator built in the United States since the Grandpa's Knob experiment more than thirty years ago. The project will study the performance, operating characteristics, and economics of wind systems for the generation of commercial electric power. The metal rotor blades will be located on the downwind side of the tower. The generator is expected to produce 100 KWe output at 40 rpm at a wind speed of 18 mph, and should generate 180,000 KWH/year in the form of 460 volt, 3-phase, 60 cycle ac power.[139]

If both wind generators and solar-thermal or photovoltaic generators become commercially viable, using both in a utility power network will be a definite advantage since the availability of sunlight seldom coincides with the availability of wind. Winds are often most vigorous in cloudy weather when photovoltaic array outputs would be minimal and solar-thermal power systems would not work at all. On the other hand, on clear, calm days plants using sunlight would function best and wind plants not at all. If a large number of generators using sunlight and wind are dispersed over a large geographical area but connected to the *same power grid*, storage requirements would be minimal.

For small applications, solar heat and wind power systems complement each other well, because sunlight is most useful at present for providing heat and wind generators for producing power. Wind can now be used to meet electrical power needs while solar energy is used for heating and cooling.

Chapter 15

SOLAR ENERGY TOMORROW

Over millions of years plants covered the earth, converting the energy of sunlight into living tissue, some of which was buried in the depths of the earth to produce deposits of coal, oil, and natural gas. During the past few decades man has found many valuable uses for these complex chemical substances, manufacturing from them plastics, textiles, fertilizer and the various end products of the petrochemical industry. Each decade sees increasing uses for these products. Coal, oil, and gas are non-renewable natural resources which will certainly be of great value to future generations, as they are to ours.

However, man has found another use for these valuable chemicals from the earth—a use other than the creation of the products that add so much to our standard of living. That is to burn them. To burn them in huge and ever increasing quantities to power the machines of society and provide heat. They are being burned at such an incredible rate that in a few short decades the world reserves of natural gas many be depleted, decades later the oil will be gone, and in a century or two the world will also be without coal. Undoubtedly successive generations after that time will decry the excesses of the present generation in selfishly destroying these valuable resources without regard for the welfare of their descendants.

It should now be clear to the reader that the rapid depletion of non-renewable fossil resources need not continue, since it is now or soon will be technically and economically feasible to supply all of man's energy needs from the most abundant energy source of all, the sun. Sunlight is not only inexhaustible, but it is the only energy source which is completely non-polluting. The land area required to provide all our energy is a small fraction of the land area required to produce our food, and the land *best* suited for collecting solar energy (rooftops and deserts) is the land *least* suited for other purposes. It is time for the United States, which led the world in the development of atomic energy and putting men on the moon, to mount an equally massive effort to usher in the Solar Age. A massive federal effort by our country can offer the world the technology for the economical utilization of solar energy in all its varied forms —photovoltaic, direct solar-thermal, renewable fuels, ocean thermal, and wind. Then we can conserve our valuable non-renewable fossil resources for future generations to enjoy, and we can all live in a world of abundant energy without pollution.

REFERENCES

1. "Reference Energy Systems and Resource Data", *Associated Universities, Inc.*, AET-8, April 1972.
2. Cherry, William R., "Harnessing Solar Energy: The Potential", *Astronautics & Aeronautics*, p. 30–36, Aug. 1973.
3. Rose, David J., "Energy Policy in the U.S.", *Scientific American*, Vol. 230, No. 1, pp. 20–29, January 1974.
4. Gaucher, Leon P., "The Solar Era: Part I—The Practical Promise", *Mechanical Engineering*, pp. 9–12, August 1972.
5. Gambs, Gerard C., *The Energy Crisis in the United States*, published by Ford, Bacon & David, Inc., March 1973.
6. Ritchings, Frank A., "Trends in Energy Needs", *Mechanical Engineering*, pp. 18–23, August 1972.
7. Nixon, Richard M., "Energy Policy", Message to the Congress Announcing Executive Actions and Proposing Enactment of Bills to Provide for Energy Needs, *Presidential Documents*, Vol. 9, No. 16, pp. 389–406, April 18, 1973.
8. Dupree, W. G., Jr. and West, James A., "United States Energy Through the Year 2000", *Dept. of Interior*, U.S. Govt. Printing Office, 1973.
9. Schulman, Fred, "Technology, The Energy Crisis, and our Standard of Living", *Mechanical Engineering*, pp. 16–23, September 1973.
10. "Toward a National Energy Policy", *Environmental Science & Technology*, Vol. 7, No. 5, p. 392–397, May 1973.
11. Altman, M., Telkes, M. and Wolf, M., "The Energy Resources and Electric Power Situation in the United States", *Energy Conversion*, Vol. 12, pp. 63–64, 1972.
12. Gambs, G. C. and Rauth, A. A., "The Energy Crisis", *Chemical Engineering*, p. 56–68, May 31, 1971.
13. Donovan, Paul and Woodward, William, "An Assessment of Solar Energy As a National Energy Resource", *NSF/NASA Solar Energy Panel* (Univ. of Maryland), December 1972.

14. "Patterns of Energy Consumption in the United States", *Stanford Research Institute Report,* January 1972.
15. Löf, G. O. G. and Tybout, R. A., "Cost of House Heating with Solar Energy", *Solar Energy,* Vol. 14, pp. 253–278, 1973.
16. "Energy Research and Development Space Technology", *Hearings of the Committee on Science and Astronautics, U.S. House of Representatives,* U.S. Government Printing Office, Washington, D.C., May 1973.
17. Böer, K. W., "Future Large Scale Terrestrial Use of Solar Energy", *Proceedings of the NASA Conference on Solar and Chemical Power,* QC-603, p. 145–148, 1972.
18. Löf, G. O. G., Duffie, J. A. and Smith, C. O., "World Distribution of Solar Energy", *Solar Energy,* Vol. 10, No. 1, pp. 27–37, 1966.
19. Angstrom, A., "Solar and Terrestrial Radiation," *Journal of the Royal Meterological Society,* Vol. 50, No. 121, 1924.
20. *Transactions of the NSF/NOAA Solar Energy Data Workshop,* Washington, D.C., Nov. 29–30, 1973, In press.
21. Kakabaev, A. and Golaev, M., "Experimental Study of the Thermal Output of Some Simple Solar Heater Designs", *Geliotekhnika,* Vol. 7, No. 2, pp. 41–46, 1971 (Russian).
22. Lorsch, Harold G., "Performance of Flat Plate Collectors", *Proceedings of the NSF/RANN Solar Heating and Cooling for Buildings Workshop,* Washington, D.C., NSF/RANN–73–004, pp. 1–14, July 1973.
23. Streed, Elmer R., "Some Design Considerations for Flat Plate Collectors", *Proceedings of the NSF/RANN Solar Heating and Cooling for Buildings Workshop,* NSF/RANN–73–004, pp. 26–35, July 1973.
24. Merriam, Marshall, E., "Materials Technology for Flat Plate Steam Generation", *Proceedings of the NSF/RANN Solar Heating and Cooling for Buildings Workshop,* NSF/RANN–73–004, pp. 17–20, July 1973.
25. Böer, Karl W., "Solar Collectors of the University of Delaware Solar House Project", *Proceedings of the NSF/RANN Solar Heating and Cooling for Buildings Workshop,* NSF/RANN–73–004, pp. 15–16, July 1973.
26. Edmondson, W. B., ed., "Breakthrough in Selective Coatings", *Solar Energy Digest,* Vol. 2, No. 2, p. 1–2, February 1974.
27. Teplyakov, D. I., "Analytic Determination of the Optical Characteristics of Paraboloidal Solar Energy Concentrators," *Geliotekhnika,* Vol. 7, No. 5, pp. 21–33, 1971 (Russian).
28. Giutronich, J. E., "The Design of Solar Concentrators Using Toroidal Spherical, or Flat Components", *Solar Energy,* Vol. 7, No. 4, pp. 162–166, 1963.
29. Eibling, James A., "A Survey of Solar Collectors", *Proceedings of the*

NSF/RANN Solar Heating and Cooling for Buildings Workshop, NSF/RANN–73–004, pp. 47–51, July 1973.

30. Lidorenko, N. S., Nabiullin, F. Ka, Landsman, A. P., Tarnizhevskii, B. V., Gertsik, E. M. and Shul'meister, L. F., "An Experimental Solar Power Plant", *Geliotekhnika,* Vol. 1, No. 3, pp. 5–9, 1965 (Russian).

31. Fairbanks, J. W. and Morse, F. H., "Passive Solar Array Orientation Devices For Terrestrial Application", *Solar Energy,* Vol. 14, pp. 67–69, 1972.

32. Gunter, Carl, "The Utilization of Solar Heat for Industrial Purposes by Means of a New Plane Mirror Reflector", *Scientific American,* May 26, 1906.

33. Steward, W. Gene, "A Concentrating Solar Energy System Employing a Stationary Spherical Mirror and Movable Collector", *Proceedings of the NSF/RANN Solar Heating and Cooling for Buildings Workshop,* NSF/RANN–73–004, pp. 17–20, July 1973.

34. Russell, J. L., DePlomb, E. P. and Ravinder, R. K., "Principles of the Fixed Mirror Solar Concentrator", *Solar Energy,* submitted for publication.

35. "Solar Energy Research—A Multi-disciplinary Approach", *Staff Report of the Committee on Science and Astronautics of the House of Representatives,* December 1972.

36. Weingart, J. M. and Schoen, Richard, "Project SAGE—An Attempt to Catalyze Commercialization of Gas Supplemented Solar Water Heating Systems for New Apartments in Southern California", *Proceedings of the NSF/RANN Solar Heating and Cooling for Buildings Workshop,* NSF/RANN–73–004, pp. 75–91, July 1973.

37. Tybout, R. A. and Löf, G. O. G., "Solar House Heating", *Natural Resources Journal,* Vol. 10, No. 2, p. 268–326, April 1970.

38. Lorsch, Harold G., "Solar Heating/Cooling Projects at the University of Pennsylvania", *Proceedings of the NSF/RANN Solar Heating and Cooling for Buildings Workshop,* NSF/RANN–73–004, pp. 194–210, July 1973.

39. Burke, J. C., "Massachusetts Audubon Society Solar Building", *Proceedings of the NSF/RANN Solar Heating and Cooling for Buildings Workshop,* NSF/RANN–73–004, pp. 214–217, July 1973.

40. Thomason, H. E. and Thomason, H. J. L., Jr., "Solar Houses/Heating and Cooling Progress Report", *Solar Energy,* Vol. 15, pp. 27–39, 1973.

41. Thomason, H. E., "Solar Houses and Solar House Models", Edmund Scientific Company, Barrington, N. J., 1972.

42. Close, D. J., "Solar Air Heaters", *Solar Energy,* Vol. 7, pp. 117–124, 1963.

43. Satcunanthan, Sand Deonarine, S., "A Two-Pass Solar Air Heater", *Solar Energy,* Vol. 15, pp. 41–49, 1973.

44. Telkes, Maria, "Energy Storage Media", *Proceedings of the NSF/*

RANN Solar Heating and Cooling for Buildings Workshop, NSF/RANN–73–004, pp. 57–59, July 1973.

45. Hay, H. R., and Yellott, J. I., "A Naturally Air-Conditioned Building", *Mechanical Engineering*, Vol. 92, No. 1, pp. 19–25, Jan. 1970.

46. Hay, H. R., "Energy Technology and Solarchitecture", *Mechanical Engineering*, pp. 18–22, Nov. 1973.

47. Hay, Harold R., "Evaluation of Proved Natural Radiation Flux Heating and Cooling", *Proceedings of the NSF/RANN Solar Heating and Cooling for Buildings* Workshop, NSF/RANN–73–004, pp. 185–187, July 1973.

48. Noguchi, Tetsuo, "Recent Developments in Solar Energy Research Application in Japan", *Solar Energy*, Vol. 15, pp. 179–187, 1973.

49. Kusuda, T., "Solar Water Heating in Japan", *Proceedings of the Solar Heating and Cooling for Buildings Workshop*, NSF/RANN–73–004, pp. 141–147, July 1973.

50. Garg, H. P., "Design and Performance of a Large-Size Solar Water Heater", *Solar Energy*, Vol. 14, pp. 303–312, 1973.

51. Garg, H. P. and Gupto, C. L., "Design Data for Direct Solar Utilization Devices—1, System Data". *Journal of the Institute of Engineers* (India), (Special Issue), Vol. 48, p. 461, 1968.

52. Liu, B. Y. H. and Jordan, R. C., "Predicting Long-Term Average Performance of Flat Plate Solar Energy Collectors", *Solar Energy*, Vol. 7, p. 53, 1963.

53. Garg, H. P. and Gupta, C. L., "Design of Flat-Plate Solar Collectors for India", *J. Inst. Engrs.*, (India), Vol. 47, p. 382, 1967.

54. Gupta, C. L. and Garg, H. P., "System Design in Solar Water Heaters with Natural Circulation", *Solar Energy*, Vol. 12, pp. 163–182, 1968.

55. Sobotka, R., "Economic Aspects of Commercially Produced Solar Water Heaters", *Solar Energy*, Vol. 10, No. 1, pp. 9–14, 1966.

56. Malik, M. A. S., "Solar Water Heating in South Africa", *Solar Energy*, Vol. 12, p. 395–397, 1969.

57. Daniels, Farrington, "Direct Use of the Sun's Energy", *American Scientist*, Vol. 55, No. 1, pp. 29–30, 1967.

58. Farben, E. A., Flanigan, F. M., Lopez, L. and Politka, R. W., "University of Florida Solar Air-Conditioning System", *Solar Energy*, Vol. 10, pp. 91–99, 1966.

59. Teagen, W. P., "A Solar Powered Heating/Cooling System with the Air Conditioning Unit Driven by an Organic Rankine Cycle Engine", *Proceedings of the NSF/RANN Solar Heating and Cooling for Buildings Workshop*, NSF/RANN 73–004, pp. 107–111, July 1973.

60. Löf, George, "Solar Cooling Design and Cost Study", *Proceedings of the NSF/RANN Solar Heating and Cooling for Buildings Workshop*, NSF/RANN–73–004, pp. 119–125, July 1973.

61. Baum, V. A., Aparase, A. R. and Garf, B. A., "High-Power Solar Installations", *Solar Energy*, Vol. 1, No. 2, pp. 6–13, 1957.

62. Edlin, F. E., "Worldwide Progress in Solar Energy", *Proceedings of the Intersociety Energy Conversion Engineering Conference* (IECEC), pp. 92–97, New York, 1968.

63. Hildebrandt, A. F. and Vant-Hull, L. L., "Large Scale Utilization of Solar Energy", *Energy Research and Development*, Hearings of the House Committee on Science and Astronautics, pp. 499–505, U.S. Govt. Printing Office, 1972.

64. Trombe, F., "Solar Furnaces and Their Applications", *Solar Energy*, Vol. 1, No. 2, pp. 9–15, 1957.

65. Walton, J. D., personal communication.

66. Meinel, A. B. and Meinel, M. P., "Energy Research and Development", *Hearings of the House Committee on Science and Astronautics*, U.S. Govt. Printing Office, pp. 583–585, 1972.

67. Russell, John L., Jr., "Investigation of a Central Station Solar Power Plant", *Proceedings of the Solar Thermal Conversion Workshop*, Washington, D.C., January 1973; (also published as *Gulf General Atomic Report* No. Gulf-GA-A12759, August 31, 1973).

68. Oman, H. and Bishop, C. J., "A Look at Solar Power for Seattle", *Proc. IECEC*, pp. 360–65, August 1973.

69. Meinel, A. B., "A Joint United States-Mexico Solar Power and Water Facility Project", Optical Sciences Center, University of Arizona, April 1971.

70. Ralph, E. L., "A Commercial Solar Cell Array Design", *Solar Energy*, Vol. 14, pp. 279–286, 1973.

71. Currin, C. G., Ling, K. S., Ralph, E. L., Smith, W. A. and Stirn, R. J., "Feasibility of Low Cost Silicon Solar Cells", *Proceedings of the 9th IEEE Photovoltaic Specialists Conference*, Silver Spring, Md., May 2–4, 1972.

72. Riel, R. K., "Large Area Solar Cells Prepared on Silicon Sheet", *Proceedings of the 17th Annual Power Sources Conference*, Atlantic City, N.J., May 1963.

73. Cherry, W. R., *Proceedings 13th Annual Power Sources Conference*, Atlantic City, N.J., pp. 62–66, May 1959.

74. Tyco Laboratories, *Final Report* #AFCRL–66–134, 1965.

75. Currin, C. G., Ling, K. S., Ralph, E. L., Smith, W. A. and Stirn, R. J., "Feasibility of Low Cost Silicon Solar Cells", *Proceedings of the 9th IEEE Photovoltaic Specialists Conference*, Silver Spring, Md., May 2–4, 1972.

76. Eckert, J. A., Kelley, B. P., Willis, R. W. and Berman, E., "Direct Conversion of Solar Energy on Earth, Now", *Proc. IECEC*, pp. 372–5, 1973.

77. Homer, Lloyd, *NSF/NOAA Solar Energy Data Workshop*, Washington, D.C., Nov. 1973.

78. Rink, J. E. and Hewitt, J. G., Jr., "Large Terrestrial Solar Arrays", *Proceedings of the Intersociety Energy Conversion Engineering Conference*, Boston, Mass., Aug. 3–5, 1971.

79. Wolf, M., *Proceedings of the 9th IEEE Photovoltaic Specialists Conference*, Silver Spring, Md., May, 1972.

80. Ralph, E. L., "A Plan to Utilize Solar Energy as an Electric Power Source", *IEEE Photovoltaic Specialists Conference*, pp. 326–330, 1970.

81. Pfeiffer, C., Schoffer, P., Spars, B. G. and Duffie, J. A., "Performance of Silicon Solar Cells at High Levels of Solar Radiation", *Transactions of the American Society of Mechanical Engineers (ASME)*, 84A, p. 33, 1962.

82. Pfeiffer, C. and Schoffer, P., "Performance of Photovoltaic Cells at High Radiation Levels", *Trans. ASME*, 85A, p. 208, 1963.

83. Beckman, W. A., Schoffer, P., Hartman, W. R., Jr. and Löf, G. O. G., "Design Consideration for a 50 Watt Photovoltaic Power System Using Concentrated Solar Energy", *Solar Energy*, Vol. 10, No. 3, pp. 132–136, 1966.

84. Tarnizhevskir, B. V., Savchenko, I. G., and Rodichev, B. Y., "Results of an Investigation of a Solar Battery Power Plant", *Geliotekhnika*, Vol. 2, No. 2, pp. 25–30, 1966 (Russian).

85. "Solar Energy", *Moscow News*, No. 2, January 1974.

86. Farber, E. A., "Solar Energy, Its Conversion and Utilization", *Solar Energy*, Vol. 14, pp. 243–252, 1973.

87. Schaeper, H. R. A. and Farber, E. A., "The Solar Era: Part 4—The University of Florida Electric", *Mechanical Engineering*, pp. 18–24, November 1972.

88. Backus, C. E., "A Solar-Electric Residential Power System", *Proc. IECEC*, 1972.

89. Rabenhorst, D. W., "Superflywheel", *Proceedings of the NSF/RANN Solar Heating and Cooling for Buildings Workshop*, NSF–RANN–73–004, pp. 60–68, July 1973.

90. Loferski, J. J., "Large Scale Solar Power Via the Photoelectric Effect", *Mechanical Engineering*, December, 1973, pp. 28–32.

91. Pope, R. B., *et al.*, "A Combination of Solar Energy and the Total Energy Concept—The Solar Community", *Proc. 8th IECEC*, pp. 304–311, 1973.

92. Schimmel, W. P., Jr., "A Vector Analysis of the Solar Energy Reflection Process", *Annual Meeting of the U.S. Section of the International Solar Energy Society*, Cleveland, Ohio, October 1973.

93. Pope, R. B. and Schimmel, W. P., Jr., "An Analysis of Linear Focused

Collectors for Solar Power", *Proc. 8th IECEC*, Philadelphia, Penn., Aug. 1973.

94. Achenbach, P. R. and Cable, J. B., "Site Analysis for the Applications of Total Energy Systems to Housing Developments", *7th IECEC*, San Diego, Ca., Sept. 1972.

95. Sakvrai, T., Osamu, K., Koro, S. and Koji, I., "Construction of a Large Solar Furnace", *Solar Energy*, Vol. 8, No. 4, 1964.

96. Akyurt, M. and Selcuk, M. K., "A Solar Drier Supplemented with Auxiliary Heating Systems for Continuous Operation", *Solar Energy*, Vol. 14, pp. 313–320, 1973.

97. Bevill, V. D. and Brandt, H., "A Solar Energy Collector for Heating Air", *Solar Energy*, Vol. 12, pp. 19–29, 1968.

98. Khanna, M. L., "Design Data for Heating Air by Means of Heat Exchanger with Reservior, Under Free Convection Conditions, For Utilization of Solar Energy", *Solar Energy*, Vol. 12, pp. 447–456, 1969.

99. Hay, Harold R., "The Solar Era: Part 3—Solar Radiation: Some Implications and Adaptations", *Mechanical Engineering*, pp. 24–29, October 1972.

100. Morse, R. N. and Read, W. R. W., "A Rational Basis for the Engineering Development of a Solar Still", *Solar Energy*, Vol. 12, pp. 5–17, 1968.

101. Lewand, T. A., "Description of a Large Solar Distillation Plant in the West Indies", *Solar Energy*, Vol. 12, pp. 509–512, 1969.

102. Hay, Harold R., "New-Roofs for Hot Dry Regions", *Ekistics*, Vol. 31, pp. 158–164, 1971.

103. Talbert, S. G., *et al.*, Manual on Solar Distillation of Saline Water, *Progress Report No. 546*, U.S. Dept. of Interior, Office of Saline Water, 1970.

104. Achilov, B., *et al.*, "Investigation of an Industrial-Type Solar Still", *Geliotekhnika*, Vol. 7, No. 2, pp. 33–36, 1971 (Russian).

105. Achilov, B. M., "Comparative Tests on Large Solar Stills in the Fields of Kzyl-kum in the Uzbek SSR", *Geliotekhnika*, Vol. 7, No. 5, pp. 86–89, 1971 (Russian).

106. Baum, V. A. and Bairamov, R., "Heat and Mass Transfer Processes in Solar Stills of the Hotbox Type", *Solar Energy*, Vol. 8, No. 3, pp. 78–82, 1964.

107. Annaev, A., Bairamov, R., and Rybakova, L. E., "Effect of Wind Speed and Direction on the Output of a Solar Still", *Geliotekhnika*, Vol. 7, No. 4, pp. 33–37, 1971 (Russian).

108. Szego, G. C., Fox, J. A., and Eaton, D. R., "The Energy Plantation", Paper no. 729168, *Proc. IECEC*, pp. 113–4, September 1972.

109. Woodmill, G. W., "The Energy Cycle of the Biosphere", *Scientific American*, Vol. 233, No. 3, p. 70, September 1970.

110. DeBeni, G. and Marchetti, C., "Hydrogen, Key to the Energy Market", *Eurospectra*, Vol. 9, No. 2, p. 46, 1970.

111. Marchetti, C., "Hydrogen and Energy", *Chemical Economy and Engineering Review*, pp. 7–25, January, 1973.

112. D'Arsonval, J., *Revue Scientifique*, Vol. 17, September 1881.

113. Claude, Georges, "Power From the Tropical Seas", *Mechanical Engineering*, Vol. 52, No. 12, pp. 1039–1044, December 1930.

114. Roe, Ralph C. and Othmer, Donald F., "Controlled Flash Evaporation", *Mechanical Engineering*, Vol. 93, No. 5, pp. 27–31, 1971.

115. *Proceedings of the Solar Sea Power Plant Conference and Workshop*, sponsored by the National Science Foundation (RANN), Carnegie-Mellon Univ., Pittsburgh, Pa., June, 1973.

116. Anderson, J. H. and Anderson, J. H., Jr., "Thermal Power from Seawater", *Mechanical Engineering*, Vol. 88, No. 4, pp. 41–46, April 1966.

117. Zener, Clarence, "Solar Sea Power", *Physics Today*, pp. 48–53, January 1973.

118. McGowan, J. G., Connell, J. W., Ambs, L. L. and Goss, W. P., "Conceptual Design of a Rankine Cycle Powered by the Ocean Thermal Difference", *Proc. IECEC*, paper no. 739120, pp. 420–27, August 1973.

119. Anderson, J. H., "The Sea Plant—A Source of Power, Water and Food Without Pollution", *Solar Energy*, Vol. 14, pp. 287–300, 1973.

120. Barnes, S., "Mining Marine Minerals", *Machine Design*, April, 1968.

121. Banks, Ridgway, personal communication, January 1974.

122. a) Glaser, P. E., "The Future of Power from the Sun", *Proc. IECEC*, pp. 98–103, 1968, and b) "Power From the Sun: Its Future", *Science*, Vol. 162, pp. 857–861, Nov. 22, 1968.

123. Glaser, P. E., "A New View of Solar Energy", *Proc. IECEC*, paper no. 719002, pp. 1–4, 1971.

124. Glaser, P. E., Maynard, O. E., Mockovciak, J. and Ralph, E. L. "Feasibility Study of a Satellite Solar Power Station", *NASA Contractor Report CR-2357*, February 1974.

125. Glaser, Peter E., "Solar Power Via Satellite", *Astronautics and Aeronautics*, pp. 60–68, August 1973.

126. Brown, William C., "Adapting Microwave Techniques to Help Solve Future Energy Problems", *Proceedings of the IEEE International Microwave Symposium*, pp. 189–191, June 1973.

127. Patha, J. T. and Woodcock, G. R., "Feasibility of Large-Scale Orbital Solar/Thermal Power Generation", *Proc. IECEC*, pp. 312–319, 1973.

128. Brown, William C., "Satellite Power Stations: A New Source of Energy?", *IEEE Spectrum*, pp. 38–47, March 1973.

129. Mockovciak, John, Jr., "A Systems Engineering Overview of the Satellite Power Station", *Proc. 7th IECEC*, paper no. 739111, pp. 712–19, Sept. 1972.

130. "Glaser, P. E., Maynard, O. E., Mackovciak, J. Jr. and Ralph, E. L.,

"Feasibility Study of a Satellite Solar Power Station", *NASA Contractor Report CR-2357*, February 1974.

131. Bergey, Karl H., "Wind Power Demonstration and Siting Problems." *Wind Energy Conversion Systems Workshop Proceedings*, NSF/RA/W–73–006, ed. by J. M. Savino, pp. 41–45, December 1973.

132. Wilcox, Carl, "Motion Picture History of the Erection and Operation of the Smith-Putnam Wind Generator", *Wind Energy Conversion Systems Workshop*, NSF/RA/W–73–006, ed. by J. M. Savino, pp. 8–10, December 1973.

133. Smith, B. E., "Smith-Putnam Wind Turbine Experiment", *Wind Energy Conversion Systems Workshop Proceedings*, NSF/RA/W–73–006, ed. by J. M. Savino, pp. 5–7, December 1973.

134. Thomas, Percy H., "Electric Power from the Wind", Federal Power Commission, Washington, D.C., March 1945.

135. Hutter, Ulrich, "Past Developments of Large Wind Generators in Europe", *Wind Energy Conversion Systems Workshop Proceedings*, NSF/RA/W–73–006, pp. 19–22, December 1973.

136. Tompkin, J., "Introduction to Voigt's Wind Power Plant", *Wind Energy Conversion Systems Workshop Proceedings*, NSF/RA/W–73–006, ed. by J. M. Savino, pp. 23–26, December 1973.

137. Clews, Henry M., "Wind Power Systems for Individual Applications", *Wind Energy Conversion Systems Workshop Proceedings*, NSF/RA/W–73–006, pp. 164–69, December 1973.

138. Noel, John M., "French Wind Generator Systems", *Wind Energy Conversion System Workshop Proceedings*, NSF/RA/W–73–006, pp. 186–196, December 1973.

139. Edmondson, William B., "A 100 KW Windmill Generator", *Solar Energy Digest*, Vol. 2, No. 4, p. 4, April 1974.

GLOSSARY

Absorption cooling. Refrigeration or air conditioning achieved by an absorption-desorption process that can utilize solar heat to produce a cooling effect.

Absorptivity. The ratio of the incident radiant energy absorbed by a surface to the total radiant energy falling on the surface.

Albedo. The ratio of the light reflected by a surface to the light falling on it.

Ambient temperature. Prevailing temperature outside a building.

Anaerobic fermentation. Fermentation process caused by bacteria in the absence of oxygen.

Bio-conversion. Use of sunlight to grow plants with subsequent use of the plants to provide energy.

Brayton cycle. Power plant using a gas turbine to drive a compressor and produce power. A gas is compressed, then heated, then expanded through a turbine, then cooled. The turbine produces more power than is needed to drive the compressor.

British Thermal Unit (BTU). A unit of energy which is equal to the amount of heat required to raise the temperature of a pound of water one degree Fahrenheit.

Capital cost. The cost of construction, including design costs, land costs, and other costs necessary to build a facility. Does not include operating costs.

Capture efficiency of plants. The ratio of the energy absorbed and converted into tissue by plants to the total solar energy falling on the plants. This energy, usually about 3% or less of the total incident solar energy, can be released when the plants are burned.

Collector efficiency. The ratio of the energy collected by a solar collector to the radiant energy incident on the collector.

Concentration ratio (concentration factor). Ratio of radiant energy intensity at the hot spot of a focusing collector to the intensity of unconcentrated direct sunshine at the collector site.

Convective heat transfer. Transfer of heat by the circulation of a liquid or gas.

Degree day (DD). One day with the average ambient temperature one degree colder than 65°F. For example, if the average temperature is 55°F for 3 days, the number of degree days is (65–55) times 3, or 30.

Diffuse insolation. Sunlight scattered by atmospheric particulates that arrives from a direction other than the direction of direct sunlight. The blue color of the sky is an example of diffuse solar radiation.

Direct conversion. Conversion of sunlight directly into electric power, instead of collecting sunlight as heat and using the heat to produce power. Solar cells are direct conversion devices.

Direct insolation. Sunlight arriving at a location that has not been scattered, also referred to as direct beam radiation.

Dynamic conversion. The collection of sunlight to heat a fluid that operates an engine to produce power. An example is the use of concentrated sunlight to boil water and operate a steam engine to produce power.

Electrolysis. The use of an electric current to produce hydrogen and oxygen from water.

Equilibrium temperature. The temperature of a device or fluid under steady operating conditions.

Faceted concentrator. A focusing collector using many flat reflecting elements to concentrate sunlight at a point or along a line.

Fossil fuel. Coal, oil or natural gas.

Fuel cell. A device, somewhat like a battery, that uses a chemical reaction to produce electricity directly, such as the reaction of hydrogen and oxygen to produce electric power with water as a product.

Geosynchronous satellite (synchronous satellite). An artificial satellite in a synchronous orbit 22,300 miles from the earth that can remain continuously above the same spot on the earth, since the period of the orbit is 24 hours.

Heat of fusion. The heat released when a liquid becomes a solid (freezes), equal to the heat absorbed when a solid melts—often tabulated in units of BTU/pound.

Heat transfer fluid. A liquid or gas that transfers heat from a solar collector to its point of use.

Heliostat. An electro-optical-mechanical device that orients a mirror so that sunlight is reflected from the mirror in a fixed specific direction, regardless of the sun's position in the sky.

Hot spot. The location on a focusing collector at which the concentrated sunlight is focused and the highest temperatures are produced. If the heat is to be collected, a heat exchanger is located at the hot spot and a heat transfer fluid flowing through the heat exchanger is heated.

Incidence angle. The angle between the direction of the sun and the perpendicular to the surface on which sunlight is falling.

Infrared radiation. Thermal radiation or light with wavelengths longer than 0.7 microns. Invisible to the naked eye, the heat radiated by objects at less than 1000°F is almost entirely infrared radiation.

Insolation. Sunlight, or solar radiation, including ultraviolet, visible and infrared radiation from the sun. Total insolation includes both direct and diffuse insolation.

KWe. Kilowatt of electric power.

KWH. Kilowatt-hour.

KWt. Kilowatt of thermal (heat) energy.

Linear concentrator. A solar concentrator which focuses sunlight along a line, such as the parabolic trough concentrator (Figure 6) and the fixed-mirror concentrator (Figure 27).

MBTU. Million BTU's.

Micron. A millionth of a meter, or micro-meter, a common unit for measuring the wavelength of light. Ultraviolet light has wavelengths less than 0.4 microns, visible light covers the wavelength range of 0.4 to 0.7 microns, and infrared radiation has wavelengths longer than 0.7 microns.

Microscale data. Refers to data on insolation and weather parameters that can vary considerably over distances of a few miles, for example there may be more cloudiness and haze near a lake or in a city than a few miles away. Microscale data can be collected by satellites.

MWe. Megawatt (million watts) of electric power.

MWt. Megawatt of thermal (heat) energy.

Mill. An amount of money equal to one-tenth of a cent.

Optical coatings. Very thin coatings applied to glass or other transparent materials to increase the transmission (reduce the reflection) of sunlight. Coatings are also used to reflect back to the heat exchanger infrared radiation emitted from it.

Phase-change material. A material used to store heat by melting. Heat is later released for use as the material solidifies.

Photovoltaic cells (solar cells). Semiconducting devices that convert sunlight directly into electric power. The conversion process is called the *photovoltaic effect.*

Pyranometer. An instrument for measuring sunlight intensity. It usually measures total (direct plus diffuse) insolation over a broad wavelength range.

Pyrheliometer. An instrument that measures the intensity of the direct beam radiation (direct insolation) from the sun. The diffuse component is not measured.

Radian. A unit of angular measure, one radian equals 57.296 degrees. The sunshine has an angular diameter of 0.009 radians, or one-half degree.

Reflectivity (reflectance). The ratio of light reflected from a surface to the light falling on the surface. The reflectivity plus the absorptivity equals one, since the incident sunlight is either reflected or absorbed.

Selective coating. An optical coating for heat exchangers that has a high absorptivity (low reflectivity) for incident sunlight (wavelengths less than one micron) and high reflectivity (low absorptivity) for infrared heat (wavelengths greater than one micron), as shown in Figure 2. The low infrared absorptivity (low emissivity) results in reduced radiant heat loss, so the collection efficiency is improved, and higher temperatures can be achieved.

Solar concentrator. Device using lenses or reflecting surfaces to concentrate sunlight.

Solar farm. A large array of solar collectors, as shown in Figure 27, for generating large amounts of electric power.

Solar furnaces. Solar concentrators for producing very high temperatures. Installations in France, Russia and Japan produce temperatures as high as 7000°F.

Solar-thermal conversion. The collection of sunlight as heat, and the conversion of heat into electric power.

Specific heat. The amount of heat required to raise the temperature of one pound of material one degree Fahrenheit, usually measured in BTU/lb°F.

Spectral pyranometer. An instrument for measuring total insolation over a restricted wavelength range.

Specular reflection. Mirror-like reflection from a surface.

Thermal efficiency. The ratio of electric power produced by a power plant to the amount of heat supplied to the plant.

Thermochemical hydrogen production. Use of heat with a series of chemical reactions to produce hydrogen from water.

Total energy system. System for providing all energy requirements, including heat, air conditioning, and electric power.

Turbidity. Atmospheric haze.

Refractory materials. Materials that can withstand high temperatures without melting.

Vapor cycle. Method of converting heat into power by boiling a liquid, expanding the vapor through a turbine, condensing the vapor back to a liquid and pumping the liquid back to the boiler. The power output of the turbine is much greater than the power required by the pump.

Waste heat. Heat rejected by a power plant.

INDEX

University of Florida Solar House (1956) heated by solar collectors shown in foreground.[86] The flat plate solar collectors and storage tank are located on the ground (instead of on the roof) to facilitate public examination. Solar energy provides hot water, space heating, swimming pool heating, electricity for some lights and appliances and fresh water from liquid wastes. (Photo courtesy of Erich Farber, Univ. Florida.) See page 61.

Solar Water Heater. Water flows through copper tubes soldered in a sinusoidal configuration to a copper sheet (painted black to enhance solar energy absorption) located inside a sheet metal box, 4 ft x 2 ft, covered by a layer of glass. One inch of styrofoam insulation beneath the plate reduces heat loss through the bottom of the box. The 80 gal storage tank is above the top of the absorber so hot water will circulate into the tank without a pump. (Photo courtesy of Erich Farber, Univ. Florida.) See page 41.

Small Solar Still used to reclaim drinking water from household liquid wastes at the University of Florida Solar House. Water evaporates from the metal pan, condenses on the sloping glass cover and is collected as distilled water. (Photo courtesy of Erich Farber, Univ. of Florida.) See page 72.

Thomason Solar House No. 3 near Washington DC (1962).[40] It has received up to 85% of its annual heat from the sun despite cloudy weather, snow and below freezing temperatures. The solar collector on the roof also is used to heat the indoor swimming pool. In summer at night a small air conditioner chills the water in the tank, which is then used during the day to cool the house. Due to the greater efficiency of night operation, energy use and air conditioning and installation costs are much less than for conventional air conditioning systems. (Photo courtesy of Omer Henry.) See page 29.

Experimental Fixed Mirror Concentrator, built by the author and a graduate student at Georgia Tech, is composed of long narrow flat mirrors arranged on a concave cylindrical surface. The angles of the mirrors are fixed so that focal distance is twice the radius of the cylindrical surface. The focus is always sharp for sunlight of any incident direction. The heat exchanger pipe is pivoted at the center of the reference cylindrical surface to remain at the focal point as sun direction changes.[67] See page 20.

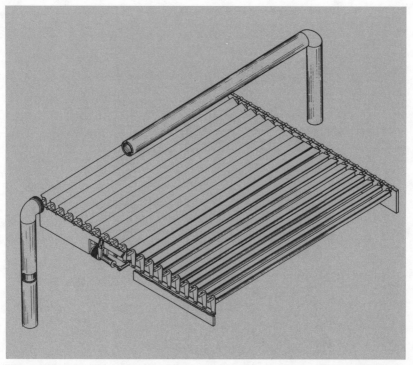

Solar Concentrator with Rotating Facets.[32] Each facet is oriented at the appropriate angle to reflect sunlight onto the heat exchanger pipe. As the sun moves a single bar rotates each facet the same amount, so that sunlight reflected from all facets remains focused on the heat exchanger. This system appears promising for collecting heat at high temperatures on building rooftops. (Drawing courtesy of Max Akridge, Georgia Institute of Technology.) See page 20.

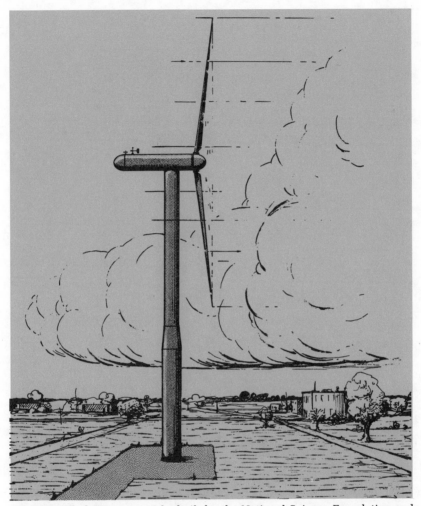

100 KWe Wind Generator to be built by the National Science Foundation and NASA at Sandusky, Ohio.[138] Rotor blades are located on the downwind side of the tower, and the alternator and transmission are contained in the enclosure on top. This is expected to generate 180,000 KWH/year in 460-volt, 3-phase, 60-cycle ac power. It will be the first large wind energy system constructed in the United States in 30 years and should be a prototype for future generators of more than one megawatt output. (Drawing courtesy of Joseph Savino, NASA.) See page 101.